Absolute Therapeutic Medical Physics Review

Malcolm Heard • Raghavendiran Boopathy
Charles R. Thomas, Jr.
Editors

Absolute Therapeutic Medical Physics Review

Questions and Detailed Answers

 Springer

Editors
Malcolm Heard
Department of Radiation Medicine
Oregon Health & Science University
Portland, OR, USA

Charles R. Thomas, Jr.
Dartmouth Cancer Center
Department of Radiation Oncology, Geisel
School of Medicine
Lebanon, NH, USA

Raghavendiran Boopathy
OU Stephenson Cancer Center
University of Oklahoma Health
Science Center
Oklahoma City, OK, USA

ISBN 978-3-031-14670-1   ISBN 978-3-031-14671-8   (eBook)
https://doi.org/10.1007/978-3-031-14671-8

This Springer imprint is published by the registered company Springer Nature Switzerland AG
The registered company address is: Gewerbestrasse 11, 6330 Cham, Switzerland

Foreword

There are few careers more important and rewarding than therapeutic medical physics. Radiation therapy (RT) is a leading treatment for cancer and involves the use of potentially lethal doses of radiation with a universally accepted absolute tolerance for delivery of 5%. RT is a medicine, so physicians are responsible for the patient, appropriateness of radiation therapy, dose prescriptions, and treatment plans. However, it is the responsibility of the medical physicist to assure that the treatment plan is technically appropriate and is accurately delivered. This is a significant and complex responsibility. Expertise in the physical principles underlying RT is critical to the medical physicist, which is why physics courses are a required prerequisite. Additionally, deep operational understanding of treatment equipment, techniques, safety issues, calculation algorithms, computer connectivity and data management, radiation safety, and general management of complex systems is also required.

My career in medical physics started on a plane. I was at the time earning my Ph.D. in Nuclear Physics from Indiana University. I was getting ready to graduate when, during a return flight from a vacation, I began chatting with the person next to me. He said that he was flying to Indianapolis to install a microtron, a device used to deliver electron beam radiation therapy. He told me that he was a medical physicist, and I asked, "What is medical physics?" He explained as much as he could, and short story long upon graduating, I joined MD Anderson as a postdoctoral fellow under the tutelage of Dr. Starkschall and Dr. Hogstrom. At the time, MD Anderson allowed their postdoctoral fellows to attend the master's in medical physics courses, and I took advantage of that to learn about diagnostic imaging, radiation therapy, and anatomy. After completing my postdoctoral fellowship, I joined Washington University under the supervision and guidance of Dr. James Purdy, Ph.D., as an instructor, a rank I maintained temporarily to effectively extend my tenure deadline. Unlike myself, Dr. Purdy was well aware of the differences in knowledge gained during academic instruction versus clinical instruction. He made me a sort of proto-resident for a year, spending 6 months in dosimetry, 3 months in brachytherapy, and 3 months as a physics trainee with the physics group. This was in 1991, 7 years before Washington University became the first accredited residency program. Dr. Purdy's commitment to clinical education has guided me since, keeping the clinical mission at the forefront even through my research and administrative careers both at Washington University and now at UCLA.

Clinical medical physics education processes have deepened and been formalized since I joined Washington University with the advent of medical physics residencies. These residencies teach the profession of medical physics, including the broad array of information and technical skills previously mentioned. Accreditation of these residency programs assures a level of quality uniformity, while formal certification by the American Board of Radiology (ABR) provides evidence that the medical physicist has been suitably educated and is qualified independently as a clinical medical physicist. ABR certification currently requires two written and one oral examination. While the knowledge gained during the education one receives during the formal degree program and residency is sufficient to pass these tests, each candidate can and should refresh their knowledge, if only to update and emphasize what they have already learned, prior to taking the exams.

It is to aid these candidates, as well as anyone refreshing their knowledge, that resources such as these are critical, and I am excited and thankful that Drs. Heard, Boopathy, and Thomas elected to gather experts throughout our community to prepare this study guide. They have done a wonderful job, and I applaud their hard work.

Radiation Oncology, UCLA   Daniel Low
Los Angeles, CA, USA

Contents

Contributors

Max Amurao Department of Radiation Oncology, Washington University in St. Louis, St. Louis, MO, USA

Arjit K. Baghwala Department of Radiation Oncology, Houston Methodist Hospital, Houston, TX, USA

Nina Bahar Radiation Oncology, St. Peter's Hospital, Albany, NY, USA

Charles Bloch Radiation Oncology, University of Washington School of Medicine, Seattle, WA, USA

Raghavendiran Boopathy Department of Radiation Oncology, OU Stephenson Cancer Center, University of Oklahoma Health Science Center, Oklahoma City, OK, USA

Erli Chen Radiation Oncology Department, Cheshire Medical Center, Keene, NH, USA

Richard J. Crilly Radiation Medicine, Oregon Health and Science University, Portland, OR, USA

Malcolm Heard Radiation Medicine, Oregon Health and Science University, Portland, OR, USA

Brandon Merz Radiation Medicine, Oregon Health and Science University, Portland, OR, USA

Susha Pillai Radiation Medicine, Oregon Health and Science University, Portland, OR, USA

Prema Rassiah-Szegedi Radiation Oncology, Huntsman Cancer Hospital, University of Utah, Salt Lake City, UT, USA

Matthew G. Rodriguez Radiation Oncology, NorthShore University Health System, Evanston, IL, USA

Vasanthan Sakthivel Department of Radiation Oncology, Cancer Specialists of North Florida, Jacksonville, FL, USA

Martin Szegedi Radiation Oncology, University of Utah, Huntsman Cancer Hospital, Salt Lake City, UT, USA

Ganesh Tharmarnadar Manipal Hospitals Dwaraka, Dwaraka, New Delhi, India

Benjamin B. Williams Radiation Oncology Department, Dartmouth Hitchcock Medical Center, Lebanon, NH, USA

Chapter 1
Conformal Radiation Therapy

Raghavendiran Boopathy and Malcolm Heard

1. In developing a 3D conformal treatment plan, a CT may be performed to define which of the following:

   A. Gross tumor volume (GTV)
   B. Conformity index (CI)
   C. Organs at risk (OR)
   D. Setup margin (SM)
   E. A and C

   Answer: The correct answer is E. The GTV is the gross demonstrable extent and location of the malignant growth that may be visualized from different imaging procedures (CT, MRI, PET). Organs at risk are normal tissues that are sensitive to radiation and may need consideration during planning. The conformity index is the ratio between the treated volume and the planning target volume (PTV). The setup margin accounts for uncertainties in patient positioning and alignment. Setup margins can vary based on many factors including beam patient immobilization, beam arrangement, and machine type [1].

R. Boopathy (✉)
Department of Radiation Oncology, OU Stephenson Cancer Center, University of Oklahoma Health Science Center, Oklahoma City, OK, USA
e-mail: raghavendiran-boopathy@ouhsc.edu

M. Heard
Radiation Medicine, Oregon Health and Science University, Portland, OR, USA
e-mail: heardma@ohsu.edu

2. Which one of the following techniques would be used to treat a chest wall patient?

    A. Tangential photon fields
    B. Accelerated partial breast irradiation (APBI)
    C. Three-field single isocenter
    D. Prone irradiation

    Answer: The correct answer is C. Tangential fields are utilized to encompass the chest wall with modifications made to include the mastectomy scar or drain sites. A half-beam block is used to field match the tangential fields and supraclavicular field, thus creating a mono-isocentric plan. The supraclavicular field is angled away from the spinal cord and encompasses the supraclavicular and axillary nodes. The use of a single isocenter minimizes setup/treatment time in addition to dose uncertainties at the match line. However, the maximum tangential field length is limited to half of the full field size, 20 cm for most linac machines. Tangential photon fields and prone irradiation are used for the treatment of intact breast. APBI treats only the lumpectomy cavity inclusive of a 1–2 cm margin [2].

3. Monitor unit (MU) calculations are performed using the dose per MU under normalization conditions for a specific photon or electron beam. What is the recommended normalization depth for an electron beam? What is the recommendation for a photon beam?

    A. dmax; 10 cm
    B. dref; dmax
    C. dref; 10 cm
    D. dmax; dmax

    Answer: The correct answer is A. For electron beams, the depth of normalization is taken to be the depth of maximum dose. For photon beams, this task group recommends that a normalization depth of 10 cm be selected. This recommendation differs from the more common approach of a normalization depth of dmax, although both systems are acceptable within the current protocol [3].

4. Matching two adjacent photon fields at the patient surface will result in overdosing of tissues at depth.

    A. True
    B. False

    Answer: The correct answer is A. Due to beam divergence, a hotspot will occur in the tissues at depth. The gap can be calculated such that the 50% isodose levels add up at depth and the overlap is minimized [4].

5. Parallel opposed beams with a field size = 10×10 cm and 100 cm SSD will be used to treat a patient with a 27 cm separation. What energies are appropriate to ensure that the dose uniformity is within 10%?

   A. Co-60
   B. 6 MV
   C. 10 MV
   D. 18 MV

   Answer: The correct answer is D. The central axis dose near the patient surface increases relative to the midpoint dose as the patient thickness increases. This is called the tissue lateral effect [5].

6. In the figures A and B shown below, which of the following is true?

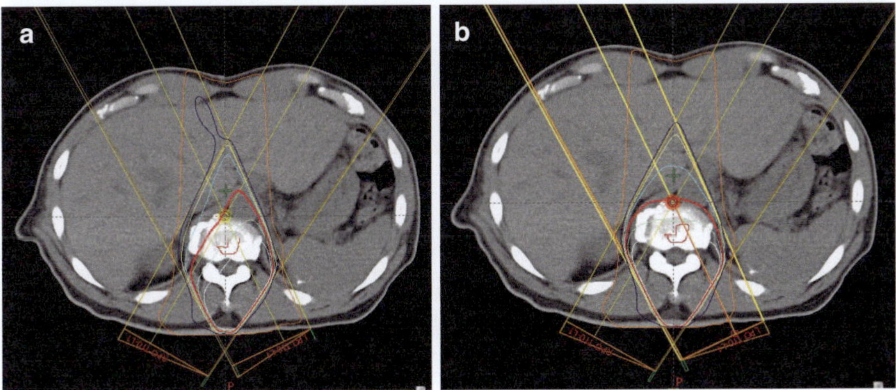

   A. CT image quality is inadequate for dose calculation.
   B. Field sizes are too big.
   C. Wedge angles are wrong.
   D. Collimator angles or wedge orientation should be changed so that thicker ends are facing together.

   Answer: The correct answer is D. Collimator angles or wedge orientation should be changed so that thicker ends are facing each other [5].

7. What could be done to improve the 100% isodose coverage of the vertebrae in the picture shown below?

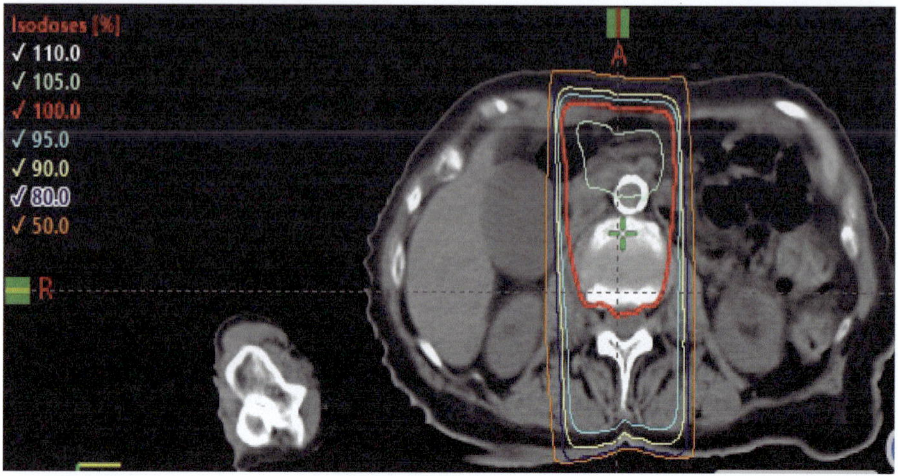

A. New HU values to be assigned.
B. Use different dose calculation algorithm.
C. Anterior/posterior beam weight to be adjusted.
D. Protons/electrons should be preferred over photons.

Answer: The answer is C. The posterior beam weight should be increased to improve the coverage of vertebrae and decrease the hotspot [6].

8. What is the treatment technique shown in the figures below?

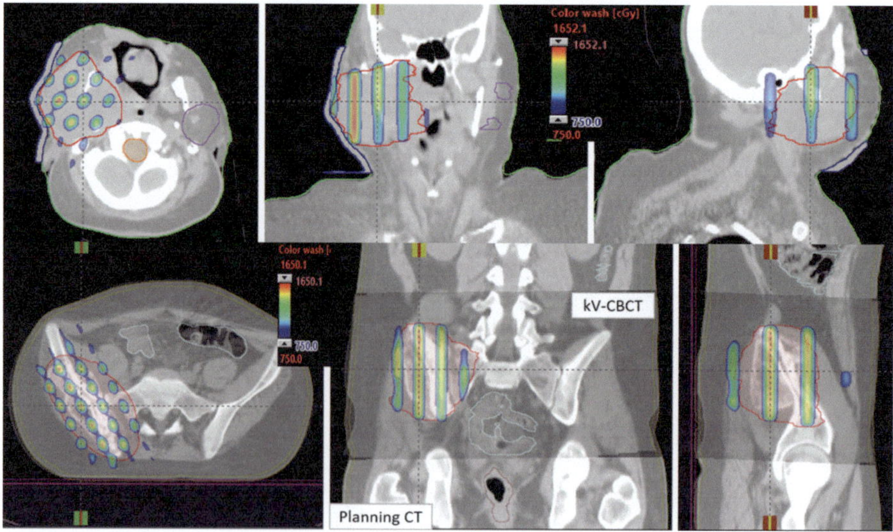

A. Spatially fractionated radiation therapy
B. Interstitial brachytherapy
C. Radium-223 dichloride therapy
D. Flash therapy

Answer: The answer is A. Grid therapy or spatially fractionated therapy: standard dose is 10–20 Gy/fraction typically used to treat palliative larger tumor [7].

9. MR images can be used for dose calculation in whole-brain treatment with 5% dose uncertainty.

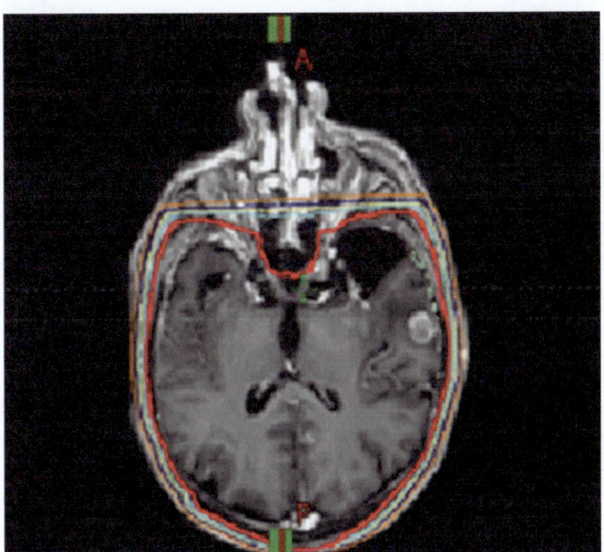

A. True
B. False

Answer: The answer is A. The dose calculation accuracy of MR-based planning for brain is within 5% compared to CT-based planning [8].

10. In the picture below, which of the following statement is correct?

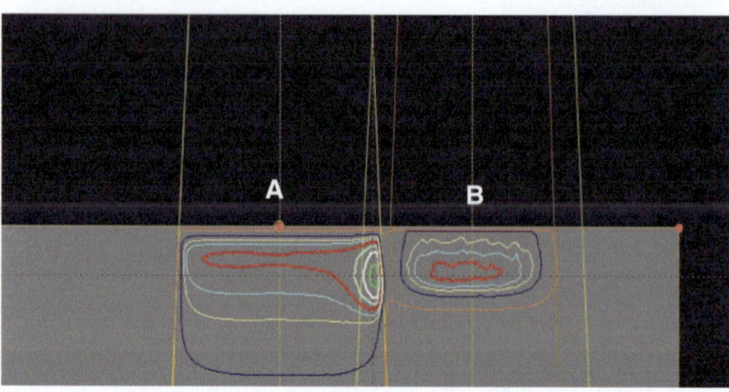

A. Field A is a photon beam and field B is an electron beam.
B. Field A is an electron beam and field B is a photon beam.
C. Both fields A and B are electron beams.
D. Both fields A and B are photon beams.

Answer: The answer is A. When an electron field is abutted at the surface with a photon field, a hotspot develops on the side of the photon field and a cold spot develops on the side of the electron field. This is caused by outscattering of electrons from the electron field [5].

References

1. Landberg T, Chavaudra J, Dobbs J, Gerard JP, Hanks G, Horiot JC, Johansson KA, Möller T, Purdy J, Suntharalingam N, Svensson H. Report 62. Journal of the ICRU. 1999;32(1). https://doi.org/10.1093/jicru/os32.1.Report62.
2. Khan, Faiz M. Treatment planning in radiation oncology. 2011.
3. Gibbons JP, Antolak JA, Followill DS, Huq MS, Klein EE, Lam KL, Palta JR, Roback DM, Reid M, Khan FM. Monitor unit calculations for external photon and electron beams: report of the AAPM therapy physics committee task group no. 71. Med Phys. 2014;41:031501.
4. Podgorsak EB. Radiation oncology physics: a handbook for teachers and students. Vienna: IAEA; 2005. p. 255–6.
5. Khan M. The physics of radiation therapy. 4th ed. Philadelphia: Lippincott Williams & Wilkins; 2009.
6. Khan FM, Potish RA. Treatment planning in radiation oncology. Baltimore: Williams & Wilkins; 1998.
7. Yan W, Khan MK, Wu X, Simone CB 2nd, Fan J, Gressen E, Zhang X, Limoli CL, Bahig H, Tubin S, Mourad WF. Spatially fractionated radiation therapy: history, present and the future. Clin Transl Radiat Oncol. 2019;22(20):30–8. https://doi.org/10.1016/j.ctro.2019.10.004.
8. Prabhakar R, Julka PK, Ganesh T, Munshi A, Joshi RC, Rath GK. Feasibility of using MRI alone for 3D radiation treatment planning in brain tumors. Jpn J Clin Oncol. 2007;37(6):405–11. https://doi.org/10.1093/jjco/hym050.

Chapter 2
IMRT/VMAT

Matthew G. Rodriguez

1. Which setting has been adjusted between the two treatment plans shown below?

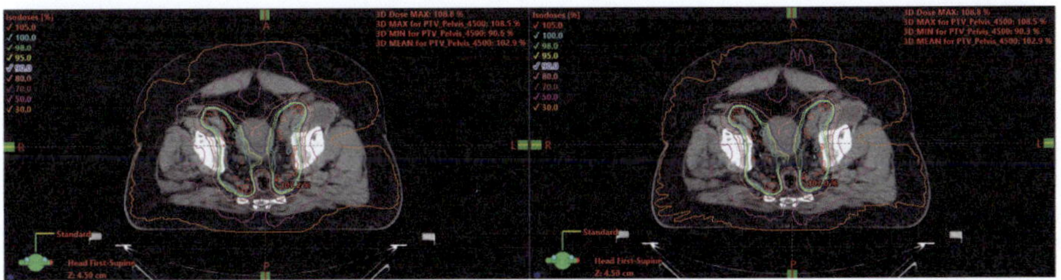

A. Dose grid size
B. Angular resolution
C. Calculation algorithm

Answer: The correct answer is B. The image on the left was calculated using an angular resolution of 2°. The image on the right was calculated using an angular resolution of 5°. The treatment plan has a control point resolution of 2°, which in the above example would correspond to 178 control points. By adjusting the angular resolution to 5°, the treatment planning system will merge adjacent control points to a total of 74 control points. Although this will significantly speed up the calculation time, it can cause an apparent ripple effect in the calculated dose, most clearly seen with the 30% isodose line in the above example [1].

M. G. Rodriguez (✉)
Radiation Oncology, NorthShore University Health System, Evanston, IL, USA

M. Heard et al. (eds.), *Absolute Therapeutic Medical Physics Review*,
https://doi.org/10.1007/978-3-031-14671-8_2

2. You have calculated an IMRT verification plan onto an array using a dose grid of 3 mm. If you adjust the dose grid to 1 mm and create a second verification plan, would your gamma pass rate increase, decrease, or stay the same?

A. Increase
B. Decrease
C. Stay the same

Answer: The correct answer is A. By switching to a finer resolution, there will be more dose points for the distance-to-agreement (DTA) component of the gamma pass rate to find a satisfactory result. When reducing the DTA tolerance value for your gamma pass rate, it is recommended that you use a fine dose grid resolution [2].

3. Which tests should be performed on a monthly basis for a linear accelerator that is performing IMRT?

A. MLC transmission and dosimetric leaf gap
B. MLC leaf travel speed and MLC leaf position accuracy
C. MLC spoke shot and light field vs. treatment field coincidence

Answer: The correct answer is B. MLC transmission, dosimetric leaf gap, spoke shot, and light field vs. treatment field coincidence should all be performed on an annual basis. The tolerance for leaf travel speed should be <0.5 cm/s from baseline. The tolerance for leaf position accuracy should be 1 mm [3].

4. What is the largest field size that should be used to verify small field percent depth dose (PDD) when performing VMAT/IMRT validation?

A. 10×10 cm^2
B. 5×5 cm^2
C. 3×3 cm^2
D. 2×2 cm^2

Answer: The correct answer is D. When verifying small field PDDs, a field size less than or equal to 2×2 cm^2 should be used on MLC shaped fields [4].

5. When using a thimble ionization chamber to perform IMRT QA, what is the most important chamber parameter to consider?

A. Central electrode composition
B. Size of sensitive volume
C. Thimble cap composition
D. Sensitivity of chamber

Answer: The correct answer is B. Volume effects of the detector used for IMRT QA can lead to significant differences between the calculated and measured dose values. For accuracy, a smaller sensitive volume is preferred to a larger one [5].

6. Below is a dose volume histogram (DVH) showing the dose to the brainstem (orange) and left parotid (pink) for two VMAT head and neck treatment plans. Both plans were optimized using the same optimization objectives. What machine parameter would explain the difference between the two treatment plans?

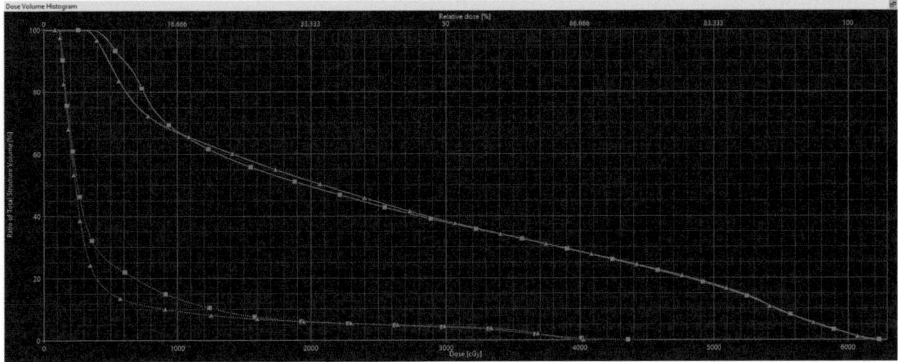

A. Dynamic jaw motion
B. Dynamic dose rate
C. Dynamic gantry rotation rate

Answer: The correct answer is A. The DVH for the plan indicated by the triangle utilized dynamic jaw motion during treatment delivery. Treatment plans that utilize this function will notice a reduction in the low-dose regions (V5 and V10) for organs at risk when compared to treatment plans delivered with static jaws [6].

7. When treating a lung cancer patient with IMRT, what would prove to be most useful in achieving a high local control rate while minimizing toxicity?

A. Respiratory gated treatment
B. 4D CT
C. Abdominal compression
D. Increased planning target volume margin

Answer: The answer is B. Respiratory gating can be beneficial; however, it requires continuous verification and can only be used on a minority of patients. Abdominal compression can drastically reduce lung tumor motion, but it does not eliminate it completely. Increasing the planning target volume margin would improve control rate but would also increase toxicity. A 4D CT allows for improved target delineation and localization. In addition, it allows for generation of a maximum intensity projection (MIP) and average intensity projection (AVG-IP) images, which can further improve dosimetric accuracy [7–9].

8. When generating a treatment plan for the breast using an inverse planned multibeam IMRT technique, what would you expect to occur relative to the conventional tangent-based technique?

A. Maximum dose to the target volume increases.
B. Maximum dose to the organs at risk increases.
C. Minimum dose to the organs at risk increases.
D. Target conformity decreases.

Answer: The answer is C. Although an inversely planned multibeam IMRT plan may have some advantages over a conventional treatment plan, it can cause a higher dose to adjacent healthy tissues [10].

9. A head and neck plan is not meeting certain dose constraints despite using the IMRT technique to achieve a sharp dose falloff. The radiation oncologist would like to reduce the PTV margin from your clinic's standard of 5–3 mm to meet the dose constraints. What should you recommend to accomplish this goal without significantly affecting the patient's local-regional control rate?

A. Utilize surface monitoring during treatment.
B. Increase the prescription dose.
C. Daily IGRT.
D. Perform a new CT simulation halfway through the course of treatment for the purpose of generating a new treatment plan.

Answer: The answer is C. Several studies have found that a PTV margin reduction from 5 mm to 3 mm will have no effect on the local control rate if the patient is treated with daily IGRT. In fact, certain clinical trials will allow for a similar reduction in PTV margin if certain IGRT credentialing is performed [11–13].

10. A patient has bilateral hip prosthesis and needs to be treated for prostate cancer. What technique should be used to reduce dosimetric uncertainty?

A. Override density of prosthesis to that of water.
B. Avoid entry through the prosthesis using static field IMRT.
C. Avoid entry through the prosthesis using VMAT with avoidance sectors.
D. Calculate the plan using a Monte Carlo-based algorithm.
E. Either B or C.

Answer: The correct answer is E. In order to minimize the dosimetric uncertainty, an attempt should be made to avoid entering through any prosthesis. Although some algorithms have shown to have improved accuracy when dealing with higher density materials, CT artifacts caused by the prosthesis can lead to further uncertainty [14].

11. As IMRT has become the standard of care in your clinic, the radiation oncologist would like to phase out sequential boosts for simultaneous integrated boosts. Previously, all head and neck plans in your clinic would consist of an initial 24 fractions, followed by a 9-fraction boost. Both plans would have a daily dose rate of 210 cGy per treatment. If the radiation oncologist would like to continue with the 33-fraction course of treatment, what doses should you recommend for the two PTVs?

A. 6600 and 5400 cGy
B. 6930 and 5961 cGy
C. 7000 and 4800 cGy
D. 6600 and 6300 cGy

Answer: The correct answer is B. This dose scheme delivered as a simultaneous integrated boost was found to have a comparable outcome to the aforementioned sequential boost plan. It should be noted that there are other possible dose schemes, but a dose per fraction greater than 160 cGy should be maintained [15, 16].

References

1. Stambaugh C, Gagneur J, Uejo A, Clouser E, Ezzell G. Improvements in treatment planning calculations motivated by tightening IMRT QA tolerances. J Appl Clin Med Phys. 2019;20:250–7.
2. Chun M, Kim J, Oh D, Wu H, Park J. Effect of dose grid resolution on the results of patient-specific quality assurance for intensity-modulated radiation therapy and volumetric modulated arc therapy. Int J Radiat Res. 2020;18(3):521–30.
3. Klein E, Hanley J, Bayouth J, Yin FF, Simon W, Dresser S, Serago C, Aguirre F, Ma L, Arjomandy B, Liu C, Sandin C, Holmes T. Task group 142 report: quality assurance of medical accelerators. Med Phys. 2009;36:4197–212.
4. Smilowitz JB, Das IJ, Feygelman V, Fraass BA, Kry SF, Marshall IR, Mihailidis DN, Ouhib Z, Ritter T, Snyder MG, Fairobent L. AAPM medical physics practice guideline 5.A.: commissioning and QA of treatment planning dose calculations—megavoltage photon and electron beams. J Appl Clin Med Phys. 2015;16:14–34.
5. Laub WU, Wong T. The volume effect of detectors in the dosimetry of small fields used in IMRT. Med Phys. 2003;30:341–7.
6. Thongsawad S, Khamfongkhruea C, Tannanonta C. Dosimetric effect of jaw tracking in volumetric-modulated arc therapy. J Med Phys. 2018;43(1):52–7.
7. Borm KJ, Oechsner M, Wiegandt M, et al. Moving targets in 4D-CTs versus MIP and AIP: comparison of patients data to phantom data. BMC Cancer. 2018;760
8. Huang L, Park K, Boike T, Lee P, Papiez L, Solberg T, Ding C, Timmerman RD. A study on the dosimetric accuracy of treatment planning for stereotactic body radiation therapy of lung cancer using average and maximum intensity projection images. Radiother Oncol. 2010;96(1):48–54.
9. Slotman BJ, Lagerwaard FJ, Senan S. 4D imaging for target definition in stereotactic radiotherapy for lung cancer. Acta Oncol. 2006;45(7):966–72.
10. Haciislamoglu E, Colak F, Canyilmaz E, Zengin AY, Yilmaz AH, Yoney A, Bahat Z. The choice of multi-beam IMRT for whole breast radiotherapy in early-stage right breast cancer. Springerplus. 2016;5(1):688.

11. Chen AM, Farwell DG, Luu Q, Donald PJ, Perks J, Purdy JA. Evaluation of the planning target volume in the treatment of head and neck cancer with intensity-modulated radiotherapy: what is the appropriate expansion margin in the setting of daily image guidance? Int J Radiat Oncol Biol Phys. 2011;81(4):943–9.
12. Nabavizadeh N, Elliott DA, Chen Y, Kusano AS, Mitin T, Thomas CR Jr, Holland JM. Image guided radiation therapy (IGRT) practice patterns and IGRT's impact on workflow and treatment planning: results from a National Survey of American Society for Radiation Oncology members. Int J Radiat Oncol Biol Phys. 2016;94(4):850–7.
13. Roger L. Phase III trial of observation versus irradiation for a gross totally resected grade II meningioma. NRG-BN003. ClinicalTrials.gov NCT # 03180268, 2017.
14. Prabhakar R, Kumar M, Cheruliyil S, Jayakumar S, Balasubramanian S, Cramb J. Volumetric modulated arc therapy for prostate cancer patients with hip prosthesis. Rep Pract Oncol Radiother. 2013;18(4):209–13.
15. Maciejewski B, Withers HR, Taylor JM, Hliniak A. Dose fractionation and regeneration in radiotherapy for cancer of the oral cavity and oropharynx: tumor dose-response and repopulation. Int J Radiat Oncol Biol Phys. 1989;16(3):831–43.
16. Vlacich G, Stavas MJ, Pendyala P, et al. A comparative analysis between sequential boost and integrated boost intensity-modulated radiation therapy with concurrent chemotherapy for locally-advanced head and neck cancer. Radiat Oncol. 2017;12:13.

Chapter 3
Stereotactic Procedures

Matthew G. Rodriguez and Malcolm Heard

1. What plan metric is used to find the optimal prescription isodose line for radio-surgery treatment?

    A. Prescription dose
    B. Target conformality
    C. Gradient index
    D. Maximum point dose

    Answer: The correct answer is C. The gradient index is used to quantify the dose falloff for a treatment plan and define an optimal prescription isodose line. Choosing an optimized isodose line will reduce the toxicity to normal tissue. Larger target volumes are typically associated with a higher prescription isodose line. It is not recommended that you use prescription isodose line drop below 50% as it can lead to large heterogeneity in the dose distribution [1].

2. What MRI imaging technique is most appropriate to visualize a metastasis for the purpose of treating with radiosurgery?

    A. Axial 3D spoiled gradient + contrast with <1.5 mm slices
    B. T2 3D volume with <1.5 mm slices
    C. Axial 3D spoiled gradient + contrast with <3 mm slices
    D. T2 3D volume with <3 mm slices

    Answer: The correct answer is A. All images used for radiosurgery should have a slice thickness less than 1.5 mm. In order to visualize a brain metastasis,

M. G. Rodriguez (✉)
Radiation Oncology, NorthShore University Health System, Evanston, IL, USA

M. Heard
Radiation Medicine, Oregon Health and Science University, Portland, OR, USA
e-mail: heardma@ohsu.edu

M. Heard et al. (eds.), *Absolute Therapeutic Medical Physics Review*,
https://doi.org/10.1007/978-3-031-14671-8_3

an axial 3D spoiled gradient image with contrast is needed. However, depending on the disease being treated, other images may be appropriate. A few of the most common radiosurgery diseases treated and their recommended MRI imaging technique(s) are listed below:

Metastasis: axial 3D spoiled gradient + contrast
Acoustic neuroma: axial 3D spoiled gradient + contrast and T2 3D volume
Meningioma: axial 3D spoiled gradient + contrast and T2 3D volume
Trigeminal neuralgia: axial 3D spoiled gradient + contrast and T2 3D volume
Arteriovenous malformation: axial 3D spoiled gradient + contrast and digital subtraction angiography [2]

3. How is a target localized when treating using Gamma Knife?

   A. Implanted markers
   B. Cranial alignment
   C. Localizer box
   D. A or C

   Answer: The correct answer is C. Depending on the treatment planning image being acquired, an MRI or CT localizer box is attached to the head frame. The treatment planning system will map landmarks within the localizer box and generate a coordinate system for the planning image relative to the head frame. This allows for submillimeter accuracy of target localization [3].

4. What type of imaging is required to generate a treatment plan for a Gamma Knife?

   A. MRI only
   B. CT only
   C. MRI and CT
   D. A or B

   Answer: The correct answer is D. A Gamma Knife plan can be generated with either an MRI or a CT. However, most Gamma Knife plans are created with only the use of an MRI. Due to limitations of an MRI (spatial distortion, no electron density material table, etc.), the Gamma Knife planning system requires accurate measurements of the surface of the head and then considers all material within the head to be water equivalent. Dose is calculated with a series of TMR calculations [4–6].

5. What radioisotope is used in a Gamma Knife?

   A. Ir-192
   B. Cs-137
   C. Co-60
   D. I-125

   Answer: The correct answer is C. Depending on the model, a Gamma Knife unit has 192–201 Co-60 sources, each with an initial activity of ~30 Ci. As a result, a Gamma Knife unit has a total initial activity of ~6000 Ci. Since the Gamma Knife uses radioactive material, its use is dictated by NRC 10 CFR 35 (or its respective state regulation if the unit is located in an agreement state). In addition, because of its high activity, the Gamma Knife is considered to contain a Category 1 amount of Co-60. This means that additional security steps must be taken as dictated by 10 CFR 37 [7, 8].

6. What chamber calibration factor is used to calibrate a Gamma Knife unit?

   A. Exposure/air kerma calibration coefficient
   B. Absorbed dose to water calibration coefficient
   C. Air kerma strength calibration coefficient
   D. A or B

   Answer: The correct answer is B. A Gamma Knife cannot be calibrated using TG-51. Historically, a modified version of TG-21 was used, which uses exposure/air kerma instead of the absorbed dose to water used by TG-51. However, TG-178 was recently released entitled Recommendations on the Practice of Calibration, Dosimetry, and Quality Assurance for Gamma Stereotactic Radiosurgery. In that document, it is recommended that the absorbed dose to water calibration coefficient for the ionization chamber be used to calibrate the machine [9–11].

7. When defining an internal target volume for a lung lesion on a 4D CT, which CT dataset should be used to ensure that the internal target volume is completely encompassing the delineated target?

   A. The average intensity projection (AIP)
   B. The maximum intensity projection (MIP)
   C. The minimum intensity projection (Min-IP)
   D. Each phase of the 4D CT

   Answer: The correct answer is D. Although the MIP can reduce the time needed to define an internal target volume, exclusively using the MIP can underestimate the target size. As a result, if a MIP is used to contour the target, it is strongly recommended that each phase of the 4D CT be used to verify the contour [12].

8. When should abdominal compression be applied for lung SBRT cases?

   A. When tumor motion is greater than 2 cm
   B. For all upper lung lesions
   C. When tumor motion is greater than 1 cm
   D. For all lower lung lesions

   Answer: The correct answer is C. Abdominal compression was found to be useful in patients with tumor motion greater than 1 cm [13].

9. What factor has a substantial dose-response relationship for local control for non-small cell lung cancer SBRT?

   A. Tumor size
   B. PTVmean BED10 > 125 Gy
   C. Treatment duration
   D. PTV minimum dose

Answer: The correct answer is B. A 2-year local recurrence rate of 4% was found for PTVmean BED10 > 125 Gy versus 17% for <125 Gy [14].

10. How many days should be allowed to pass between a patient's MRI and SRS treatment before local control rates will be affected?

    A. 7 days
    B. 14 days
    C. 21 days
    D. 28 days

    Answer: The correct answer is B. A delay greater than 14 days may cause inadequate treatment coverage to a continually growing tumor that will reduce local control rates [15].

11. Historically, in SRS, an MRI has been used for target delineation and a CT has been used for treatment planning and daily image matching. In order to mitigate the need for a target margin, how should the MRI and CT be fused together?

    A. Skull-based fusion
    B. Tumor-based fusion
    C. Soft-tissue-based fusion
    D. Marker-based fusion

    Answer: The correct answer is C. The difference between skull-based and soft-tissue-based fusion can deviate by >2 mm. In order to not have to add a target margin, a soft-tissue match should be performed between the MRI and CT. This can only be achieved if both the MRI and CT are performed with IV contrast [16].

12. When performing radiosurgery to multiple lesions with a single isocenter, what factor was found to most affect target coverage, especially for small lesions?

    A. Translational errors
    B. Lesion distance from isocenter
    C. Rotational errors
    D. Number of lesions

    Answer: The correct answer is C. Rotational errors were found to significantly compromise target coverage [17].

13. What are the most important independent predictors of brain necrosis for patients treated with SRS?

    A. Brain volume irradiated at 10 and 12 Gy
    B. Location of lesion(s) and prescription dose
    C. Brain volume irradiated at 18 Gy
    D. The number of lesions and prescription dose

    Answer: The correct answer is A. The brain volumes irradiated at 10 and 12 Gy are the most important independent predictors for brain necrosis in SRS. Risk increases significantly as the volume of V10 and V12 Gy increases [18].

14. When performing 3-fraction SRT, what dose to the brain volume is the most significant prognosis factor for radionecrosis?

  A. 15 Gy
  B. 18 Gy
  C. 21 Gy
  D. 24 Gy

   Answer: The correct answer is B. The brain volume receiving 18 Gy in 3 fractions was found to be the most significant prognosis factor for radionecrosis. The incidence of radionecrosis was 5% for V18 Gy < 30.2 cm^3 [19].

15. In a patient with no prior radiation therapy, what maximum point doses to the optic pathway in 1, 3, and 5 fractions were associated with a <1% risk of radiation-induced optic neuropathy?

  A. 8 Gy, 15.3 Gy, and 23 Gy
  B. 9 Gy, 20 Gy, and 31 Gy
  C. 10 Gy, 20 Gy, and 25 Gy
  D. 12 Gy, 18 Gy, and 31 Gy

   Answer: The correct answer is C. The risk of optic neuropathy is 1% for 10 Gy in 1 fraction, 20 Gy in 3 fractions, and 25 Gy in 5 fractions for patients with no prior radiation therapy. A greater than 10% risk is associated with these same doses for patients with prior radiation therapy to the optic apparatus [20].

16. When performing QA on an SRS plan using portal dosimetry, what is the acceptable passing rate?

  A. 3%/1 mm > 90%
  B. 3%/2 mm > 95%
  C. 2%/2 mm > 95%
  D. 4%/1 mm > 90%

   Answer: The correct answer is A. What is considered an acceptable passing rate is dependent upon the QA equipment being used. For Delta4 array and portal dosimetry, the passing rate is 3%/1 mm, while it is 3%/2 mm for an ArcCHECK and 3%/1 mm for an SRS MapCheck. Despite the gamma analysis passing criteria, the number of points exceeding or meeting these criteria should be greater than 90%. For SBRT cases, the passing rate will change to 4%/1 mm, 3%/2 mm, and 2%/1 mm for Delta4/portal dosimetry, ArcCHECK, and SRS MapCheck, respectively [21].

17. Which of the following statements about end-to-end (E2E) testing is true:

  A. E2E testing can be used to refine site-specific standard operating procedures and verify that clinical team members understand their task.
  B. The physicist should perform each step of E2E testing to ensure that the step is performed correctly.
  C. E2E testing should only be performed prior to implementation of an SRS/SBRT program.
  D. E2E is not needed to assess spatial targeting accuracy of motion management systems.

Answer: The correct answer is A. An E2E test should be considered as a "dry run" of the entire treatment process; therefore, each step should be performed by the staff member who will perform the step when the program is implemented clinically. E2E should be performed pre-implementation in addition to being done on a periodic basis (annually or quarterly depending on the machine type). E2E testing should include each aspect of the treatment process, including motion management [22].

18. During the delivery of an SRS treatment, real-time tracking can be performed using the following technology:

    A. CBCT
    B. Orthogonal X-ray imaging
    C. Optical surface guidance
    D. MV ports

    Answer: The correct answer is C. Optical surface guidance has been used in patient positioning and monitoring for several treatment sites. This includes intracranial treatment with the use of open-faced mask [23, 24].

19. For SBRT cases, the CT scan should extend ___ cm superior and inferior beyond the treatment field borders, and a calculation grid of __ mm should be used.

    A. 2 cm; 1 mm
    B. 15 cm; 1.5 mm
    C. 10 cm; 3 mm
    D. 5 cm; 2 mm

    Answer: The correct answer is B. According to AAPM TG-101, the scan length should extend 5–10 cm beyond the treatment field borders, up to 15 cm for noncoplanar beam arrangements. The scan should also include any relevant critical structures. The report also recommends an isotropic grid size of 2 mm or finer [25].

20. Which secondary imaging modality is utilized to contour the spinal cord for a spine SRS case?

    A. PET
    B. T2-weighted MRI
    C. CT myelogram
    D. B or C

    Answer: The correct answer is D. The MRI is commonly used except in cases where significant metal artifacts are present. In these cases, the cord can be delineated on a CT myelogram [26].

References

1. Zhao B, Jin JY, Wen N, Huang Y, Siddiqui MS, Chetty IJ, Ryu S. Prescription to 50-75% isodose line may be optimum for linear accelerator based radiosurgery of cranial lesions. J Radiosurg SBRT. 2014;3(2):139–47.
2. Leksell radiosurgery. (2019). Switzerland: S. Karger AG.
3. Bednarz G, Downes MB, Corn BW, Curran WJ, Goldman HW. Evaluation of the spatial accuracy of magnetic resonance imaging-based stereotactic target localization for gamma knife radiosurgery of functional disorders. Neurosurgery. 1999;45(5):1156.
4. Cheung JY, Yu KN, Yu CP, Ho RT. Dose distributions at extreme irradiation depths of gamma knife radiosurgery: EGS4 Monte Carlo calculations. Appl Radiat Isot. 2001;54(3):461–5.
5. Ertl A, Saringer W, Heimberger K, Kindl P. Quality assurance for the Leksell gamma unit: considering magnetic resonance image-distortion and delineation failure in the targeting of the internal auditory canal. Med Phys. 1999;26:166–70.
6. Zhang P, Dean D, Wu QJ, Sibata C. Fast verification of Gamma Knife treatment plans. J Appl Clin Med Phys. 2000;1(4):158–64.
7. 10 CFR 35. Medical use of byproduct material. https://www.nrc.gov/reading-rm/doc-collections/cfr/part035/index.html
8. 10 CFR 37. Physical protection of category 1 and category 2 quantities of radioactive material. https://www.nrc.gov/reading-rm/doc-collections/cfr/part037/index.html
9. McDonald D, Yount C, Koch N, Ashenafi M, Peng J, Vanek K. Calibration of the Gamma Knife Perfexion using TG-21 and the solid water Leksell dosimetry phantom. Med Phys. 2011;38(3):1685–93.
10. Committee TGRT. Recommendations on the practice of calibration, dosimetry, and quality assurance for gamma stereotactic radiosurgery. Med Phys. 2021;
11. Committee TGRT. A protocol for the determination of absorbed dose from high-energy photon and electron beams. Med Phys. 1983;10:741–71.
12. Borm KJ, Oechsner M, Wiegandt M, Hofmeister A, Combs SE, Duma MN. Moving targets in 4D-CTs versus MIP and AIP: comparison of patients data to phantom data. BMC Cancer. 2018;18(1):760.
13. Molitoris JK, Diwanji T, Snider JW 3rd, Mossahebi S, Samanta S, Onyeuku N, Mohindra P, Choi JI, Simone CB 2nd. Optimizing immobilization, margins, and imaging for lung stereotactic body radiation therapy. Transl Lung Cancer Res. 2019;8(1):24–31.
14. Kestin L, Grills I, Guckenberger M, Belderbos J, Hope AJ, Werner-Wasik M, Sonke JJ, Bissonnette JP, Xiao Y, Yan D. Elekta Lung Research Group. Dose-response relationship with clinical outcome for lung stereotactic body radiotherapy (SBRT) delivered via online image guidance. Radiother Oncol. 2014;110(3):499–504.
15. Seymour ZA, Fogh SE, Westcott SK, Braunstein S, Larson DA, Barani IJ, Nakamura J, Sneed PK. Interval from imaging to treatment delivery in the radiation surgery age: how long is too long? Int J Radiat Oncol Biol Phys. 2015;93(1):126–32.
16. Xu Q, et al. Assessment of brain tumor displacements after skull based registration: a CT/MRI fusion study. Austin J Radiat Oncol Cancer. 2015;1:1011.
17. Roper J, Chanyavanich V, Betzel G, Switchenko J, Dhabaan A. Single-isocenter multiple-target stereotactic radiosurgery: risk of compromised coverage. Int J Radiat Oncol Biol Phys. 2015;93(3):540–6.
18. Minniti G, Clarke E, Lanzetta G, Osti MF, Trasimeni G, Bozzao A, Romano A, Enrici RM. Stereotactic radiosurgery for brain metastases: analysis of outcome and risk of brain radionecrosis. Radiat Oncol. 2011;15(6):48.

19. Minniti G, Scaringi C, Paolini S, Lanzetta G, Romano A, Cicone F, Osti M, Enrici RM, Esposito V. Single-fraction versus multifraction (3 × 9 gy) stereotactic radiosurgery for large (>2 cm) brain metastases: a comparative analysis of local control and risk of radiation-induced brain necrosis. Int J Radiat Oncol Biol Phys. 2016;95(4):1142–8.
20. Milano MT, Grimm J, Soltys SG, Yorke E, Moiseenko V, Tomé WA, Sahgal A, Xue J, Ma L, Solberg TD, Kirkpatrick JP, Constine LS, Flickinger JC, Marks LB, El Naqa I. Single- and multi-fraction stereotactic radiosurgery dose tolerances of the optic pathways. Int J Radiat Oncol Biol Phys. 2021;110(1):87–99.
21. Xia Y, Adamson J, Zlateva Y, Giles W. Application of TG-218 action limits to SRS and SBRT pre-treatment patient specific QA. J Radiosurg SBRT. 2020;7(2):135–47.
22. Halvorsen PH, Cirino E, Das IJ, Garrett JA, Yang J, Yin FF, Fairobent LA. AAPM-RSS medical physics practice guideline 9.a. for SRS-SBRT. J Appl Clin Med Phys. 2017;18:10–21. https://doi.org/10.1002/acm2.12146.
23. Covington EL, Fiveash JB, Wu X, Brezovich I, Willey CD, Riley K, Popple RA. Optical surface guidance for submillimeter monitoring of patient position during frameless stereotactic radiotherapy. J Appl Clin Med Phys. 2019;20:91–8. https://doi.org/10.1002/acm2.12611.
24. Covington EL, Stanley DN, Fiveash JB, Thomas EM, Marcrom SR, Bredel M, Willey CD, Riley KO, Popple RA. Surface guided imaging during stereotactic radiosurgery with automated delivery. J Appl Clin Med Phys. 2020;21:90–5. https://doi.org/10.1002/acm2.13066.
25. Benedict SH, Yenice KM, Followill D, Galvin JM, Hinson W, Kavanagh B, Keall P, Lovelock M, Meeks S, Papiez L, Purdie T, Sadagopan R, Schell MC, Salter B, Schlesinger DJ, Shiu AS, Solberg T, Song DY, Stieber V, Timmerman R, Tomè WA, Verellen D, Wang L, Yin F-F. Stereotactic body radiation therapy: The report of AAPM Task Group 101. Med Phys. 2010;37:4078–101. https://doi.org/10.1118/1.3438081.
26. Schmitt D, Blanck O, Gauer T, et al. Technological quality requirements for stereotactic radiotherapy. Strahlenther Onkol. 2020;196:421–43. https://doi.org/10.1007/s00066-020-01583-2.

Chapter 4
Brachytherapy

Arjit K. Baghwala and Richard J. Crilly

1. When reviewing a sequence of fractions from a past single-plan HDR Ir-192 treatment, what parameter should they have that is constant?

    A. Treatment time
    B. Source activity
    C. Source activity *treatment time
    D. Source activity/treatment time

Answer: C. Optimized square and rectangular single-plane implants have also been systematically analyzed and dose indices developed in support of dose calculation verification. Define the dose index I as $I = (D \cdot A)/(S \cdot T)$, where D is the dose in cGy at a distance d from the plane of the implant, A is the area of the implant in cm^2, S is the source activity in Ci, and T is the total dwell time in seconds [1].

2. If a source is stuck out of the after-loader during a QA procedure who is responsible for removing the radiation from the room?

    A. Person doing the QA
    B. Nearest authorized medical physicist
    C. Radiation safety officer
    D. Device vendor

A. K. Baghwala (✉)
Department of Radiation Oncology, Houston Methodist Hospital, Houston, TX, USA
e-mail: abaghwala@houstonmethodist.org

R. J. Crilly
Radiation Medicine, Oregon Health and Science University, Portland, OR, USA
e-mail: crilly@ohsu.edu

M. Heard et al. (eds.), *Absolute Therapeutic Medical Physics Review*,
https://doi.org/10.1007/978-3-031-14671-8_4

Answer: D. Retrieval of just the source or source element generally may require special retrieval equipment and radiation precautions and should not be considered as part of the normal emergency action sequence involving radiation therapy operators and radiation oncologists. Normally, the vendor will assume responsibility for source retrieval [2].

3. Ir-192 decay is ~__%/day.

    A. 0.5%
    B. 1%
    C. 3%
    D. 5%

Answer: B. The half-life for Ir-192 is 73.83 days [3].

4. How long can you stand 1 m away from a 10 Ci HDR Ir-192 source before you reach the annual allowed dose for a radiation worker?

    A. 2.75 min
    B. 25 min
    C. 66 min
    D. 2.2 h

Answer: C [1, 4].

5. If the distance in question 4 is reduced to 20 cm, how much is the allowed exposure time lessened?

    A. ½
    B. 1/5
    C. 1/10
    D. 1/25

Answer: D [1].

6. Which of the following isotopes is not used for permanent prostate implants?

    A. I-125
    B. Ir-192
    C. Cs-131
    D. Pd-103

Answer: B [5]. For an effective permanent prostate seed implant, the radioactive source should have short half-life and low photon energy. The half-lives and average photon energies are 60 days and 27 keV for ^{125}I, 17 days and 23 keV for ^{103}Pd, and 9.7 days and 29 keV for ^{131}Cs, respectively.

7. Match the isotope to the energy* associated with its use in therapy.

1.	I-125	(a)	662 keV
2.	Ir-192	(b)	1250 keV
3.	Pd-103	(c)	28 keV
4.	Co-60	(d)	21 keV
5.	Cs-131	(e)	380 keV
6.	Au-198	(f)	412 keV
7.	Cs-137	(g)	29.4 keV

[a] The energy of the radiation effective to treatment may be from a short-lived daughter

Answer: 1 = c, 2 = e, 3 = d, 4 = b, 5 = g, 6 = f, 7 = a [6–8].

8. The original COMS protocol indicated that a minimum dose of 100 Gy using I-125-loaded plaques was sufficient to achieve equivalence with enucleation in treating choroidal melanoma. Upon the adoption of TG-43 calculation methods in 1996, the given dose from the trial was revised as having been:

    A. 110 Gy
    B. 90 Gy
    C. 85 Gy
    D. 76 Gy

Answer: C. In 1996, the dose prescription of 100 Gy based on pre-TG-43 dosimetry was revised to 85 Gy following the introduction of the TG-43 formalism. This dose was prescribed to the tumor apex when the tumor apex was 5 mm and to 5 mm when the tumor apex was $\geq$5 mm. In 2003, the American Brachytherapy Society recommended prescribing to the tumor apex for all medium-sized choroidal melanomas, even those <5 mm in height [9].

9. When the inhomogeneity of the Silastic membrane and the gold plaque is properly taken into account, the dose along the central axis calculated is reduced by __% for I-125 from standard TG-43 calculations.

    A. 2
    B. 5
    C. 10
    D. 15

Answer: C. In the 1990s, Chiu-Tsao et al. and de la Zerda et al. reported thermoluminescent dosimeter (TLD) measurements and Monte Carlo (MC) radiation transport simulations of the dose distributions in an eye phantom for a single 125I and 103Pd source in a COMS plaque. These groups observed central-axis dose reductions of 10% and 16% for 125I and 103Pd, respectively, and off-axis dose reductions up to 30%. They attributed these reductions to the presence of the plaque's Silastic insert (silicone polymer seed carrier) and the gold-alloy (Modulay) backing (where the term backing includes the plaque collimating lip) [9].

10. According to the NRC, the most common form of failure in HDR is:

    A. Wrong Rx used in plan
    B. Wrong catheter length entered
    C. Source decay not taken into account
    D. Applicator failure

Answer: B [10].

11. What is the maximum suggested distance between catheters in a surface mold for the treatment of skin lesions?

    A. 5 mm
    B. 7 mm
    C. 10 mm
    D. 15 mm

Answer: C [11].

12. What is the current maximum lateral distance measurement accuracy suggested in an ultrasound system used in prostate brachytherapy?

    A. 0.5 mm
    B. 1 mm
    C. 2 mm
    D. 3 mm

Answer: B [12].

13. Which of the following implant systems results in the most uniform dose distribution across the Tx site?

    A. Quimby
    B. Manchester
    C. Paris
    D. Stockholm

Answer: B [7].

14. According to the GEC ESTRO guidelines for breast brachytherapy, the PTV is:

    A. Equal to the CTV
    B. Equal to the CTV plus a 1 cm margin allowing for 5 mm from skin surface
    C. Equal to the CTV plus a 2 cm margin allowing for 5 mm from skin surface
    D. Equal to the GTV plus a 1 cm margin allowing for 5 mm from skin surface

Answer: A [13].

15. According to GEC ESTRO guidelines for locally advanced carcinoma of the cervix, the combined dose to the PTV (D90) should be (complete or partial response to EBT):

    A. Dose >= 80 Gy
    B. EQD2 >= 80 Gy
    C. D_{equiv} >= 80 Gy
    D. A point >= 80 Gy

Answer: B [14].

16. What is the suggested imaging method for planning prostate permanent seed implants?

    A. CT
    B. MRI
    C. Ultrasound

Answer: C [15].

17. The period of time between HDR factions for interstitial multicatheter breast brachytherapy should be:

    A. Calculated depending on the dose and number of fractions
    B. 6 h
    C. 4 h
    D. 8 h

Answer: B [16].

18. Shielding for a room using 10 Ci, Ir-195 HDR source with the goal of full use (maximum patient load) will require typical shielding of:

    A. 0.5–1.2 cm Pb
    B. 1.2–3.0 cm Pb
    C. 3.0–4.0 cm Pb
    D. > 4 cm Pb

Answer: C [17].

19. By NRC regulations, who must be present during the initiation of the HDR treatment?

    A. A brachytherapy-trained therapist, an authorized medical physicist
    B. A brachytherapy-trained therapist, an authorized user
    C. An authorized user and an authorized medical physicist
    D. An authorized user and a radiation safety officer

Answer: C (2). High-dose-rate remote after-loader units require (i) an authorized user and an authorized medical physicist to be physically present during the initiation of all patient treatments involving the unit, and (ii) an authorized medical physicist and either an authorized user or a physician, under the supervision of an authorized user, who has been trained in the operation and emergency response for the unit, to be physically present during continuation of all patient treatments involving the unit [18].

20. According to NRC, which of the following constitute a medical event?

   i. The total dose differs from the prescribed dose by 10%.
  ii. The fractionated dose differs from the prescribed fractional dose by 20%.
 iii. The total dose delivered differs from the prescribed dose by 20% or more.
 iv. The fractionated dose differs from the prescribed fractional dose by 50% or more.

    A. i and ii
    B. iii and iv
    C. i and iv
    D. ii and iv

Answer: B [19].

21. Which R is not part of the 5 Rs of radiobiology?

    A. Reassessment
    B. Repair
    C. Reoxygenation
    D. Radiosensitivity
    E. Redistribution

Answer: A [20].

22. Which of the following is the correct 2D TG43 equation?

A. $\dot{D}(r, \theta) = S_k * \Lambda * \dfrac{GL(r,\theta)}{GL(ro,\theta o)} * g_L(r) * F(r, \theta)$

B. $\dot{D}(r_o, \theta) = S_k * \Lambda * \dfrac{GL(ro,\theta o)}{GL(ro,\theta o)} * g_L(r_o) * F(r_o, \theta)$

C. $\dot{D}(r) = S_k * \Lambda * (r_o/r)^2 * g_p(r)$

D. $\dot{D}(r, \theta) = S_k * \Lambda * g_L(r) * F(r, \theta)$

Answer: A [21].

23. In the above mentioned TG 43 U equation, what does r_o denote?

    A. Reference distance which is specified to be 1 cm
    B. Distance from the edge of the active source to the point of interest
    C. Polar angle specifying the point of interest $P(r,\theta)$
    D. Length of the source

Answer: A [21].

24. When do you have to perform the daily QA for remote after-loader?

    A. Week of the HDR treatment
    B. Anytime (before or after) the patient is treated
    C. The morning of procedure prior to patient treatment
    D. Every day

Answer: C [1].

25. In most clinical applications of after-loaders, what is a *reasonable* positional accuracy?

    A. 2 mm
    B. 1 cm
    C. 1 mm
    D. 2 cm

Answer: A. In most clinical applications of after-loaders, a positional accuracy of ±2 mm relative to the applicator system (not anatomical landmarks in the patient) is reasonable. [Note that for remote after-loaders, the NRC insists on a positional accuracy criterion of ±1 mm (policy and guidance directive FC 86–4)]. This more rigid standard is not realizable in a clinically meaningful sense for many applicator-source combinations [22].

26. What is the recommended calibration of the HDR well ionization/reentrant chamber?

    A. Every year
    B. Every 2 years
    C. At the time of purchase
    D. Twice a year

Answer: B [22].

27. Which of the following is true during the HDR source strength verification when analyzing the disparity between the measured and manufacturer data?

    A. No action required for disparity <3%
    B. Investigate for disparity between 3% and 5%
    C. Report to the manufacturer for disparity >5%
    D. All of the above

Answer: D. "We recommend that if the institution's verification of source strength disagrees with the manufacturer's data by more than 3%, the source of the disagreement should be investigated. We further recommend that an unresolved disparity exceeding 5% should be reported to the manufacturer" [23].

28. As per AAPM TG 40 recommendations, how many seeds would you need for calibrating a 100-seed prostate seed implant?

    A. 10
    B. 11
    C. 50
    D. 100

Answer: A. "… ribbons, we recommend the calibration of all seeds. For groupings with a large number of loose seeds, we recommend that a random sample containing at least 10% of the seeds be calibrated; for a large number of seeds in ribbons, a minimum of 10% or 2 ribbons (whichever is larger) should be calibrated" [23].

29. What is the maximum reading (Amp) for a dwell position in well chamber known as?

    A. Source strength
    B. Sweet spot
    C. Air kerma strength current

Answer: B. "… the customer's Farmer chamber is first connected to the microSelectron. After the calibration setup with the jig is completed, the point of maximum chamber response (i.e., the sweet spot) of the customer's Farmer chamber is found by stepping the Ir-192 source through the catheter, which is parallel to the long axis of the chamber, with the corrected ionization current versus dwell position of the source plotted" [24].

30. What is known as stochastic effect?

    A. Effects characterized by their probability of occurrence
    B. Minimum threshold necessary for the effect
    C. Severity of effect based on increasing dose
    D. Increase in probability with increasing dose
    E. A and D

Answer: E [25].

31. For a tandem and ovoid/ring, the American Brachytherapy Society (ABS) defines the location of point A to be ____:

    A. Depending on the case and location of tumor
    B. 2 cm superior along the tandem from the top of the ovoid/ring and 2 cm lateral on a perpendicular line from the tandem

   C. 2 cm superior along the tandem from the top of the ovoid/ring and 5 cm lateral on a perpendicular line from the tandem

   D. 2 cm superior from the cervical OS and 2 cm lateral on a perpendicular line from the tandem

Answer: B [26].

32. When calculating the administered volume of radium-223 dichloride, what are the factors to be taken into consideration?

   A. Patient's body weight
   B. Decay correction factor to correct for physical decay of Ra-223
   C. Radioactivity concentration of the product
   D. All of the above

Answer: D [27].

33. In LDR/HDR prostate, what does D_{90} mean?

   A. Dose delivered to 90% of the prostate volume
   B. Percentage of prostate volume receiving 90% of the prescribed dose
   C. Dose delivered to 90cc of the prostate volume
   D. Minimum peripheral dose to 90% of the prostate volume

Answer: A [28].

34. What is the approximate acute dose known to cause cancer?

   A. 100mSv
   B. 1mSv
   C. 5mSv
   D. 3mSv

Answer: A [29].

35. Define transport index (TI):

   A. Dose rate at 1 cm from the packaging surface exposure at 1 m from the packaging surface
   B. Exposure at 1 cm from the packaging surface
   C. Exposure at surface from the package

Answer: B [30].

36. How often should leak testing be performed on sealed sources?

   A. Source must be tested at intervals not to exceed 6 months.
   B. Every year.
   C. Every 2 years.
   D. Do not test if half-life is greater than 45 days.

Answer: A. "(c) *Test frequency*: (1) Each sealed source (except an energy compensation source (ECS)) must be tested at intervals not to exceed 6 months" [31].

37. What is the patient release criteria based on the activity of a I-125 prostate seed implant?

    A. <40 mCi
    B. <9 mCi
    C. <33 mCi
    D. <2 mSv

Answer: B [32].

38. In brachytherapy, the source strength is specified in:

    A. Air kerma strength
    B. Dose rate
    C. Activity
    D. Apparent activity

Answer: A [23].

References

1. Kubo HD, et al. HDR brachytherapy treatment delivery: TG59. Med Phys. 1998;25(4):391.
2. Kubo HD, et al. HDR brachytherapy treatment delivery: TG59. Med Phys. 1998;25(4):396.
3. Nath R, et al. Guidelines by the AAPM and CEX_ESTR on the use of innovative brachytherapy decide and applications: TG 167. Med Phys. 2016;43(6):3178–205.
4. NRC, 10 CFR 20.1201 Occupational dose limits for adults.
5. Kehwar TS. Use of Cesium-131 radioactive seeds in prostate permanent implants. J Med Phys. 2009;34(4):191–3.
6. Nath R. New directions in radionuclide sources for brachytherapy. Semin Radiat Oncol. 1993;3(4):278.
7. Suntharalingam, N., et al. "Chapter 13: brachytherapy: physical and clinical aspects", Radiation oncology physics: a handbook for teachers and students, Podgorsak EB (Tech. Ed.), IAEA Pub. Vienna, 2005.
8. Henschke UK, Lawrence DC. Cesium-131 seeds for permanent implants. Radiology. 85(6):1117.
9. Rivard M, et al. Comparison of dose calculation methods for brachytherapy of intraocular tumors. MedPhys. 2011;38(1):p306.
10. Fulkerson RK, et al. Surface brachytherapy: joint report of the AAPM and the GEC-ESTRO Task Group No. 253. Med Phys. 2020;47(10):e951.
11. Ouhib Z, et al. Aspects of dosimetry and clinical practice of skin brachytherapy: The American Brachytherapy Society working group report. Brachytherapy. 2015;14(6):840.
12. Siebert F-A, et al. GEC-ESTRO/ACROP recommendations for quality assurance of ultrasound imaging in brachytherapy. Radiother Oncol. 2020;147:51.
13. Major T, et al. Recommendations from GEC ESTRO Breast Cancer Working Group (II): target definition and target delineation for accelerated or boost partial breast irradiation using

multicatheter interstitial brachytherapy after breast conserving open cavity surgery. Radiother Oncol. 2016;118:p199.

14. Viswanathan AN, et al. American Brachytherapy Society consensus guidelines for locally advanced carcinoma of the cervix. Part II HDR. Brachytherapy. 2012;11:p47.

15. Salembier C, et al. Tumour and target volumes in permanent prostate brachytherapy: a supplement to the ESTRO/EAU/EORTC recommendations on prostate brachytherapy. Radiother Oncol. 2007;83(1):p3.

16. Strnad V, et al. ESTRO-ACROP guideline: interstitial multi-catheter breast brachytherapy as accelerated partial breast irradiation alone or as boost. Radiother Oncol. 2018;128:p411.

17. McGinley PH. Shielding techniques for radiation oncology facilities. 2nd ed. Madison: Medical Physics Publishing.

18. NRC 10 CFR 35.615.

19. NRC 35.3045 Report and notification of a medical event.

20. Steel GG, McMillan TJ, Peacock JH. The 5Rs of radiobiology. Int J Radiat Biol. 1989;56(6):1045–8. https://doi.org/10.1080/09553008914552491.

21. Rivard MJ, Coursey BM, DeWerd LA, Hanson WF, Saiful Huq M, Ibbott GS, Mitch MG, Nath R, Williamson JF. Update of AAPM Task Group No. 43 Report: a revised AAPM protocol for brachytherapy dose calculations. Med Phys. 2004;31:633–74. https://doi.org/10.1118/1.1646040.

22. Nath R, Anderson LL, Meli JA, Olch AJ, Stitt JA, Williamson JF. Code of practice for brachytherapy physics: report of the AAPM Radiation Therapy Committee Task Group No. 56. Med Phys. 1997;24:1557–98. https://doi.org/10.1118/1.597966.

23. Kutcher GJ, Coia L, Gillin M, Hanson WF, Leibel S, Morton RJ, Palta JR, Purdy JA, Reinstein LE, Svensson GK, et al. Comprehensive QA for radiation oncology: report of AAPM Radiation Therapy Committee Task Group 40. Med Phys. 1994;21(4):581–618. https://doi.org/10.1118/1.597316.

24. Chang L, et al. Ir-192 calibration in air with farmer chamber for HDR brachytherapy. J Med Biol Eng. 2016;36:145–52. https://doi.org/10.1007/s40846-016-0117-0.

25. Hall EJ. Radiobiology for the radiologist. Harper & Row/Lippincott. 1st edition 1973; 2nd Edition 1978; 3rd Edition 1988; Japanese Edition 1979; 4th Edition, JB Lippincott 1994; Fifth Edition, Lippincott Williams & Wilkins, 2000; Korean Edition 2004; Sixth edition, Hall, E.J. and Giaccia, A.J., Lippincott Williams & Wilkins, 2006. Seventh Edition 2011; Chinese (Taiwan) Edition 2013

26. Viswanathan AN, Thomadsen B. American Brachytherapy Society Cervical Cancer Recommendations Committee; American Brachytherapy Society. American Brachytherapy Society consensus guidelines for locally advanced carcinoma of the cervix. Part I: general principles. Brachytherapy. 2012;11(1):33–46. https://doi.org/10.1016/j.brachy.2011.07.003.

27. Dauer LT, et al. Radiation safety considerations for the use of ^{223}RaCl$_2$ DE in men with castration-resistant prostate cancer. Health physics vol. 2014;106(4):494–504. https://doi.org/10.1097/HP.0b013e3182a82b37.

28. Yu Y, Anderson LL, Li Z, Mellenberg DE, Nath R, Schell MC, Waterman FM, Wu A, Blasko JC. Permanent prostate seed implant brachytherapy: report of the American Association of Physicists in Medicine Task Group No. 64. Med Phys. 1999;26:2054–76. https://doi.org/10.1118/1.598721.

29. National Research Council. Health risks from exposure to low levels of ionizing radiation: BEIR VII–phase 2. Committee to assess health risks from exposure to low levels of ionizing radiation. Washington, DC: National Academies Press; 2006. https://www.nap.edu/resource/11340/beir_vii_final.pdf "The BEIR VII report defines low doses as those in the range of near zero up to about 100 mSv of low-LET radiation.

30. https://www.nrc.gov/reading-rm/basic-ref/students/for-educators/11.pdf

31. US NRC 10CFR 39.35 https://www.nrc.gov/reading-rm/doc-collections/cfr/part039/part039-0035.html

32. US nuclear regulatory commission 1556–Appendix U https://www.nrc.gov/docs/ML0833/ML083300045.pdf

Chapter 5
Electron Therapy

Ganesh Tharmarnadar

1. An electron beam is flat at its clinically useful depth, just like a photon beam. Which of the following is the major reason for achieving flatness in the electron beams?

    A. The diameter of the electron beam that exits the accelerating structure
    B. The bending magnet
    C. The primary collimator
    D. All the scattering components inside the linac head and the electron applicator

Answer: D—Diameter of the electron beam exiting the accelerating structure and the bending magnet is controlled by beam optics. It needs to be broadened by a suitable technique to make the beam clinically useful. This small pencil beam does not hit the primary collimator. The scattering foil, the X-ray jaws, and the electron applicator are the most important scattering components that flatten the electron beam [1].

2. The diameter of the electron beam leaving the bend magnet is:

    A. Greater than 12 mm
    B. 10–12 mm
    C. 5–10 mm
    D. Less than 3 mm

Answer: D—The electron beam leaving the bend magnet (or accelerator structure for straight-ahead machines) is about 3 mm in diameter [2, p. 144].

G. Tharmarnadar (✉)
Manipal Hospitals Dwaraka, Dwaraka, New Delhi, India

    33
M. Heard et al. (eds.), *Absolute Therapeutic Medical Physics Review*,
https://doi.org/10.1007/978-3-031-14671-8_5

3. All of the following statements are true about the dual-foil scattering system employed in current accelerators, *EXCEPT*:

   A. It significantly improves electron beam flatness characteristics.
   B. It reduces X-ray contamination.
   C. It reduces the practical range (R_p) of the electron beam.
   D. The first foil is made up of high-atomic-number material, and the second one is made of a low-Z composite material.

Answer: C—A dual-foil scattering system, with a few centimeters or more between the two foils, significantly improves electron beam flatness characteristics with reduced X-ray contamination. The first scatterer in the dual-foil system is made of a high-atomic-number material, and the second scatterer is made of a low-atomic-number material. The dual scattering foil system flattens the electron beam better than a single-foil system does. The X-ray contamination resulting from a dual-foil system is comparable to that from a scanning beam technique, which is obsolete now. However, the practical range of the electrons is related to the nominal energy of the beam and remains unaltered by the scattering foil [2, p. 144].

4. Electron beams are contaminated with bremsstrahlung X-rays. For a 20 MeV electron beam, generated by an accelerator with a dual-foil scattering system, the typical bremsstrahlung contamination is of the order of:

   A. >12%
   B. 10–12%
   C. 7–10%
   D. <5%

Answer: D—In a modern linear accelerator, typical X-ray contamination dose to a patient ranges from approximately 0.5% to 1% in the energy range of 6–12 MeV; 1% to 2%, from 12 to 15 MeV; and 2% to 5%, from 15 to 20 MeV [3].

5. In a dual scattering foil linac, the two scattering foils are separated by:

   A. >10 cm
   B. Few centimeters
   C. Few millimeters
   D. They are in close physical contact with each other

Answer: B—A dual-foil scattering system, with a few centimeters or more between the two foils, significantly improves electron beam flatness characteristics with reduced X-ray contamination [2, p. 144].

6. While treating with electron beams, the X-ray jaws (X1, X2, Y1, and Y2 jaws):

   A. Are fully retracted and parked at that position.
   B. Are opened to the extent that exactly matches the applicator field size.

    C. Are opened as a function of electron energy-field size combination, which is always larger than the applicator field size.

    D. User can adjust the opening of these jaws according to the final collimation required on the patient's skin.

Answer: C—Clinically, one of the most important linac parameters is the setting of the X-ray jaws. The vast majority of linacs in clinical use employ fixed electron cones. The settings of the variable X-ray jaws for a given energy-cone combination are usually fixed. The setting of the X-ray jaws has an effect on beam flatness and symmetry. The consistent setting of the X-ray jaws is such an important beam parameter that it is incorporated into the linac's dosimetry interlock chain [1, p. 130].

7. In electron beam intraoperative radiotherapy (IORT), which type of applicators are used?

    A. Conventional open-type applicators
    B. Conventional applicators, but closed-walls type
    C. Special Lucite or brass applicators
    D. Special copper applicators

Answer: C—Most irradiation techniques use special Lucite (PMMA) or brass applicators linked physically to the head of the treatment machine ("docking system") [4].

8. Which of the following linac parts are mounted on a carousel in a linac head? (A) Target. (B) Series of scattering foil. (C) Flattening filter. (D) Ion chamber.

    A. A and B only
    B. B and C only
    C. C and D only
    D. A, B, and C

Answer: B—[2, p. 143, 146, Figs. 8.4 and 8.6].

9. It is desired to treat a soft-tissue lesion by an electron beam. The lesion depth of 2.8 cm is to be covered by 90% isodose line. Which electron energy would you choose?

    A. 6 MeV
    B. 9 MeV
    C. 12 MeV
    D. 18 MeV

Answer: B—Depth of 90% isodose line (in cm) can be calculated as the energy of electron beam (in MeV) divided by 3.2 (or 3.3). This is only an approximation, which largely holds good. However, it is advisable to check the data for your linac and get the exact value.

The maximum depth of the PTV determines the beam energy; in unit density tissue, the electron energy should be at least approximately 3.0 (3.3) times the maximum depth of the PTV in centimeters to cover the PTV with the 80% (90%) relative dose [5–7].

10. The percent depth-dose at 6 cm for a 9 MeV electron beam is approximately:

   A. 80%
   B. 50%
   C. 20%
   D. <2%

Answer: D—Depths are normalized to the practical range, which is approximately given by E (MeV)/2 in centimeters of water.

In a modern linear accelerator, typical X-ray contamination dose to a patient ranges from approximately 0.5% to 1% in the energy range of 6–12 MeV; 1% to 2%, from 12 to 15 MeV; and 2% to 5%, from 15 to 20 MeV [3, p. 274 (Sect. 14.4.H) and p. 279 (Sect. 14.5.B)].

11. Electron beams passing through tissue medium would lose energy at the rate of:

   A. 4 MeV/cm
   B. 3 MeV/cm
   C. 2 MeV/cm
   D. 1 MeV/cm

Answer: C—The typical energy loss in tissue for a therapeutic electron beam, averaged over its entire range, is about 2 MeV/cm in water [7, p. 137].

12. The surface doses for a 15 MV photon beam and a 15 MeV electron beam are _____ and _____, respectively.

   A. 90% and 90%
   B. 10% and 90%
   C. 90% and 10%
   D. 10% and 10%

Answer: B—A 10×10 cm^2 field typically amounts to some 30% of the maximum dose for a cobalt beam, 15% for a 6 MV X-ray beam, and 10% for an 18 MV X-ray beam—p. 171.

The surface dose of electron beams is in the range from 75% to 95% (see Fig. 8.1 also)—p. 279 [8].

13. In the electron mode, for a given dose rate, the required beam current to be delivered to the electron window is _________ of the required beam current at the X-ray target in the X-ray mode.

   A. <1%
   B. 10–25%
   C. 25–40%
   D. 40–60%

Answer: A—For a given dose rate, the required beam current to be delivered to the electron window is less than 1% of the required beam current at the X-ray target in the X-ray mode [2, p. 25].

14. Which of the following materials has the highest mass stopping power for a clinical electron beam of any energy?

    A. Bone
    B. Water
    C. Aluminum
    D. Lead

Answer: B—The rate of energy loss per gram per square centimeter, MeV g^{-1} cm^{-2} (called the mass stopping power), is greater for low-atomic-number materials than for high-atomic-number materials. This is because high-atomic-number materials have fewer electrons per gram than lower atomic number materials and, moreover, high-atomic-number materials have a larger number of tightly bound electrons that are not available for this type of interaction [8, p. 275].

15. Which electron beam has the least surface dose?

    A. **4 MeV**
    B. 6 MeV
    C. 9 MeV
    D. 12 MeV

Answer: A—The percent surface dose for electron beams increases with electron energy [8, p. 279–280, Fig. 8.3].

16. When a beam of electrons passes through a medium, the scattering power of the material varies approximately as the square of the _______________.

    A. **Atomic number Z.**
    B. Incident angle of the electron beam.
    C. Kinetic energy of the electron beam.
    D. Field size of the beam.

Answer: A—The scattering power varies approximately as the square of the atomic number and inversely as the square of the kinetic energy [3, p. 258 (Sect. 14.1.B)].

17. What is the energy of a 20 MeV electron beam at 5 cm depth in water?

    A. 20 MeV
    B. 15 MeV
    C. 10 MeV
    D. 5 MeV

Answer: C—The typical energy loss in tissue for a therapeutic electron beam, averaged over its entire range, is about 2 MeV/cm in water [7, p. 137].

18. Which is the main "culprit" in the production of bremsstrahlung tail in electron beams?

    A. Beam steering coils surrounding the accelerating structure
    B. Bending magnet
    C. Primary collimator
    D. Scattering foils

Answer: D—The bremsstrahlung tail is due to bremsstrahlung interactions of electrons with the collimation system (scattering foils, chambers, collimator jaws, etc.) and the body tissues. In general, the X-ray contamination is least in the scanning beam type of accelerator, because the scattering foils are not used. The beam steering coils are used to steer the electron beam path along the accelerating structure, and the function of the bend magnet is to bend the electron beam. The electron beam does not much interact with the primary collimator. None of these three add to X-ray contamination in the beam [3, p. 273–274].

19. In the figure, which points will receive the maximum dose?

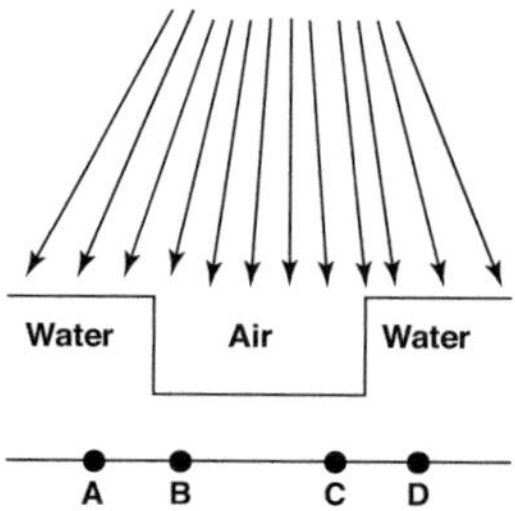

    A. Points A and B
    B. Points B and C
    C. Points C and D
    D. Points A and D

Answer: B—Sharp surface irregularities produce localized hot and cold spots in the underlying medium due to scattering. Electrons are predominantly scattered outward by steep projections and inward by steep depressions [3, p. 279 (Sect. 14.5.B)].

20. According to the recommendations of the ICRU Report 71: "*Prescribing, Recording, and Reporting Electron Beam Therapy,*"

    A. The ICRU Reference Point for reporting electron doses should always be at the center of GTV.
    B. A multiplicative correction factor of 1.05 should be used to convert the physical dose to the RBE-corrected dose.

C. Electron beam energy should be selected so that the maximum of the depth-dose curve on the beam axis ("peak dose") is reached at the center (or in the central part) of the PTV.

D. Dose is prescribed at the depth of R_{85} where the dose is 85% of the dose maximum.

Answer: C—ICRU Reference Point should be at the center (or in the central part) of PTV, not GTV (p. 51).

No weighting factor for RBE difference (relative to photons) has to be applied for the currently used electron energy range (p. 19).

In general, in electron therapy, the beam energy and the beam delivery system are adjusted so that the maximum of the depth-dose curve on the beam axis ("peak dose") is reached at the center (or in the central part) of the PTV (p. 19).

It is not the purpose of this Report (nor the role of the ICRU) to make recommendations about treatment prescription, i.e., about general approaches for prescription, dose level, beam arrangement, or other technical aspects of the treatment (p. 49) [4].

21. When an electron beam is incident at an oblique angle to the surface, all of the following can occur except:

    A. Dose maximum in the irradiated volume lies along the beam axis
    B. Shift of d_{max} toward the surface
    C. Decreased depth of penetration as measured by 80% depth
    D. Increased dose at the maximum along the beam axis

Answer: A—Beam obliquity tends to (a) increase side scatter at the depth of maximum dose (dmax), (b) shift dmax toward the surface, and (c) decrease the depth of penetration (as measured by the depth of the 80% dose). These effects are evident in Fig. 14.21.

Khan, FM and Gibbons JP (Eds). The physics of radiation therapy. Lippincott Williams & Wilkins, 2014. p 276. Section 14.5.B.

In the case of oblique incidence, the maximum dose is off-axis, and in some irradiation geometries, it may be considerably higher than the peak absorbed dose, which is defined on the beam axis [4, p. 60].

22. When using a 10 MeV electron beam, the thickness of the cutout made of low-melting-point alloy shall be:

    A. 5 mm + additional 2 mm as safety margin
    B. 10 mm + additional 2 mm as safety margin
    C. 5 mm + additional 1 mm as safety margin + 2 mm
    D. 10 mm + additional 1 mm as safety margin + 2 mm

Answer: C—A rule of thumb may be formulated: The minimum thickness of lead required for blocking in millimeters is given by the electron energy in MeV incident on lead divided by 2. Another millimeter of lead may be added as a safety margin. The required thickness of Cerrobend is approximately 20% greater than that of pure lead [3, p. 287 (Sect. 14.6.B)].

23. When a bolus is used for treating a lesion with electron beam, the SSD is set to 100 cm at:

    A. Skin surface
    B. Bolus surface
    C. Either of the two
    D. The treatment depth

Answer: A—The setup distance is measured at the level of the skin (not at the level of the bolus) [4, p. 62, Fig. 5.6].

24. At extended SSD treatments (larger than 100 cm), in general:

    A. The 90% field width is increased at extended SSDs while more dose is contributed outside of the field edge at extended SSDs.
    B. The 90% field width is decreased at extended SSDs while more dose is contributed outside of the field edge at extended SSDs.
    C. There are no changes in isodose lines until 50% isodose level below which lower isodose lines shrink.
    D. Extended SSD does not affect the shape of the isodose curves and only affects the dose rate (output).

Answer: B—In general, the 90% field width is decreased at extended SSDs while more dose is contributed outside of the field edge at extended SSDs. This change in the shape of the isodose curves is important to keep in mind so that the target is adequately covered at extended distances and for field abutment at extended SSDs [7, p. 142].

25. The dose in tissue resulting from electron beam is largely due to:

    A. Collisional losses of energy in interactions with atomic electrons in the medium.
    B. Radiative losses of energy in interactions with atomic electrons in the medium.
    C. Elastic collisions with atomic electrons.
    D. All of the above contribute to dose in approximately equal manner.

Answer: A—An electron traveling in a medium loses energy as a result of collisional and radiative processes. The magnitudes of the two effects for water and lead are shown in Fig. 14.1. The rate of energy loss of electrons of energy 1 MeV and above in water is roughly 2 MeV/cm. The rate of energy loss per centimeter in a medium due to bremsstrahlung is approximately proportional to the electron energy and to the square of the atomic number (Z2). X-ray production is more efficient for higher energy electrons and higher atomic number absorbers.

From the above, it is clear that in tissue, the most important interaction by which electrons lose their energy is collisional losses [3, p. 256–7 (Sect. 14.1.A)].

26. Modern accelerators employ which of the following technique to broaden and flatten electron beams?

    A. Single scattering foil
    B. **Dual scattering foil**
    C. Triple scattering foil
    D. Scanning beam

Answer: B—The beam collimation has been significantly improved by the introduction of the dual-foil system [3, p. 269 (Sect. 14.1.C.1)].

27. Identify the correct order in which the pencil electron beam exiting the bending magnet encounters different components before incident on the patient's/phantom's surface:

    A. Scattering foils—flattening filter—ion chamber—electron applicator
    B. Target—scattering foils—flattening filter—electron applicator
    C. Scattering foils—ion chamber—X-ray jaws—electron applicator
    D. Scattering foils—X-ray jaws—ion chamber—electron applicator

Answer: C—Fig. 8–4 on page 143 [2].

28. Which of the following would result in the least X-ray contamination (bremsstrahlung tail) of electron beam?

    A. Scanning beam technique.
    B. Single scattering foil technique.
    C. Dual scattering foil technique.
    D. All three would result in more or less equal X-ray contamination.

Answer: A—Linear accelerators employing scanning electron beams do not use scattering foils to spread the beam and thus produce the least amount of photon contamination [7, p. 138].

29. Identify the part in the linac head:

A. Carousel
B. Flattening filter
C. Single scattering foil
D. Dual scattering foil

Answer: D—Dual scattering foil.

30. For a material to be used as a bolus in electron beam therapy, it should be equivalent to tissue in:

   A. Stopping power and scattering power
   B. Physical density
   C. Stopping power and physical density
   D. Scattering power and physical density

Answer: A—Ideally, the bolus material should be equivalent to tissue in stopping power and scattering power [3, p. 283 (Sect. 14.5.D)].

31. The central axis depth-dose parameter R_{50} of an electron beam is important for:

   A. Dose prescription
   B. Electron beam quality specification
   C. Calculating the depth of D_{max} accurately
   D. Calculating the depth of R_p accurately

Answer: B—Beam quality in electron beams is specified by R_{50}, the depth in water in centimeters at which the absorbed dose falls to 50% of the maximum dose for a beam which has a field size on the phantom surface $\geq 10 \times 10$ cm^2 at an SSD of 100 cm [7].

32. According to the ICRU 71 Report, the selection of ICRU Reference Point for prescribing and reporting on the beam axis is at the:

    A. Maximum of the depth-dose curve
    B. 90% isodose level
    C. 85% isodose level
    D. 80% isodose level

Answer: A—Selecting the ICRU Reference Point for prescribing and reporting on the beam axis at the maximum of the depth-dose curve ("peak dose") appears to be a reasonable choice [4, p. 12].

33. In electron beam therapy applications, the weighting factor for RBE difference (relative to high-energy photons) to be applied for the currently used energy range is:

    A. 0.9
    B. 0.95
    C. 1.0
    D. 1.05

Answer: C—As far as relative biological effectiveness (RBE) is concerned, for the currently used electron energy range (5–35 MeV), no significant RBE difference has been observed between electron and photon beams for a wide variety of biological systems [4, p. 21].

34. In electron beam therapy, for field sizes with a diameter larger than the value of _____, the central axis depth-dose curves represent broad-beam situation.

    A. Depth of dose maximum
    B. R_{90}
    C. R_{50}
    D. **R_p**

Answer: D—Theoretical data indicate that for field sizes with a diameter larger than the value of R_p, the curves represent the broad-beam situation.

    When the field diameter is R_p or larger, broad-beam situation is reached and no further changes in central axis depth-dose curves are observed. However, below R_p, as the field size is decreased, the central axis depth-dose curves will shift toward the surface [3, p. 287; 4, p. 42].

35. Dose to patients resulting from X-ray contamination of electron beams might be of significance in:

    A. Chest wall treatment
    B. Total-skin irradiation
    C. Use of internal lead shielding
    D. Extended SSDs

Answer: B—For regular treatment field sizes used in electron beam therapy, the dose to the patient contributed by the X-ray contamination is not of much concern. However, even small amounts of X-ray contamination become critical for total skin electron irradiation such as in the treatment of mycosis fungoides [3].

36. While using a beam-shaping block, the *ideal location* to place it is:

    A. On the skin of the patient
    B. Always in the applicator as an insert
    C. Minimum 10 cm away from the skin
    D. As much away from the skin as possible

Answer: A—For electrons, the best shielding technique is to place the final collimation in contact with the patient's surface [1, p. 135 (Sect. 6.6, Fig. 6.17)].

37. When the beam-shaping block (cutout) is placed away from the skin:

    A. It will result in reduced R_p.
    B. It will produce hotspots outside the field edges.
    C. It will increase X-ray contamination in the electron beam.
    D. It will widen the penumbra and constrict the isodose lines.

Answer: D—For electrons, the best shielding technique is to place the final collimation in contact with the patient's surface. As the final collimation moves away from the surface, the width of the field's penumbra increases [1, p.135 (Sects. 6.6, 6.17)].

38. In electron beam treatments, if a patient has to be treated at extended SSD larger than the nominal SSD, the central axis percentage depth-dose will:

    A. Decrease following Mayneord factor
    B. Increase following Mayneord factor
    C. Decrease or increase depending on the energy-applicator combination
    D. Not vary significantly

Answer: D—There is very little change in the central axis percentage depth-dose with increasing SSD [7, p. 53].

39. Which of the following statements is TRUE about output factors (field size vs. output) for electron beams?

   A. They follow the same pattern as that for photons of same energy.
   B. There is no change in output with field size for electron beams and can be ignored clinically.
   C. As the field size increases, the output factor continuously decreases.
   D. Variation in output for electron beams is a function of energy-applicator-insert combination.

Answer: D—The calibration, cGy per monitor unit, for a given electron beam is dependent upon the beam energy, field (or cone) size, size and shape of the irregularly shaped field insert, and numerous other accelerator parameters.

The size of the final collimation significantly affects the output factor. While a trend across energies is demonstrated, it is difficult to draw any predictive conclusions from most measured data [1, p. 130–131].

40. Which of the following statements is TRUE about clinical electron beams?

   A. As the energy increases, surface dose decreases.
   B. Effect of field size on depth-dose is related to the range of scattered electrons in phantoms.
   C. As a thumb rule, practical range of the electron beams in centimeters is equal to one-third of the electron energy in MeV.
   D. X- and Y-jaws are set matching the electron applicator size used.

Answer: B—As the energy increases, the surface dose also increases. Practical range of the electrons in centimeters is one-half of the electron energy in MeV. The X-ray jaws are always set larger than the applicator field size.

The effect of field size on depth-dose is related to the range of the scattered electrons in phantom. In general, the field sizes with a diameter greater than one-half of the extrapolated range show minimal change in depth-dose with further increases in field size. Clinically, little change in depth-dose is realized for field sizes beyond 10 cm in diameter [1, p. 128].

41. Which of the following treatment prescription approaches does not find a mention in the ICRU Report 71?

   A. Approach based on central Reference Point
   B. Approach based on specification of a dose range within the PTV
   C. Approach based on maximum dose to the PTV
   D. Approach based on minimum dose to the PTV

Answer: C—See section 4.1.1 Prescribing the treatment—Subsections: 4.1.1.1, 4.1.1.2, and 4.1.1.3 [4, p. 49].

42. In treatments with electron beams, a sharp projection in the body contour will result in:

A. Outward scattering due to which hotspots just outside of the projection
B. Increased dose at dmax
C. Inward scattering due to which hotspots just inside of the projection
D. Decreased dose at dmax

Answer: A—Sharp surface irregularities produce localized hot and cold spots in the underlying medium due to scattering. Electrons are predominantly scattered outward by steep projections and inward by steep depressions. This can be seen in Fig. 14.23 (55). In practice, such sharp edges may be smoothed with an appropriately shaped bolus. Also, if a bolus is used to reduce beam penetration in a selected part of the field, its edges should be tapered to minimize the effect [3, p. 279–280 (Sect. 14.4.B)].

43. The use of bolus in electron beams is indicated in all of the following, *except*:

A. Increasing the skin dose
B. Compensating for missing tissues
C. Reducing X-ray dose resulting from bremsstrahlung tail
D. Lifting the isodose curves

Answer: C—Bolus is often used in electron beam therapy to (a) flatten out an irregular surface, (b) reduce the penetration of the electrons in parts of the field, and (c) increase the surface dose [3, p. 283 (Sect. 14.4.D)].

44. A surgical scar is present in the area to be treated with an electron beam. It is desired to use bolus over this thin scar to increase its surface dose. The width of the bolus shall:

A. Exactly match the thin scar
B. At least be 2 cm in width
C. At least be 5 cm in width
D. Cover the entire field size

Answer: B—If a small strip of bolus is to be used, such as to increase the surface dose to a surgical scar, then the bolus must be wide enough at least 2 cm to ensure that the dose to the skin actually is increased rather than decreased due to outscattering from the bolus and by edge effects [7].

45. CET used in electron treatments to account for inhomogeneities refers to:

A. Coefficient of equivalent thickness
B. Computed equivalent thickness
C. Calculated effective thickness
D. Characterized effective thickness

Answer: A—For large and uniform slabs, dose distribution beyond the inhomogeneity can be corrected by using the coefficient of equivalent thickness (CET) method [3, p. 280 (Sect. 14.4.C)].

46. CET for an inhomogeneity is approximately equal to:

    A. One-fourth of the electron energy (MeV)
    B. One-third of the electron energy (MeV)
    C. Square of its electron density relative to that of water
    D. Its electron density relative to that of water

Answer: D—The CET for a given material is approximately given by its electron density (electron/mL) relative to that of water [3, p. 280 (Sects. 14.5.C.1, 14.5.C.2)].

47. CET values for bone and lung are:

    A. 1.65 and 0.2–0.25
    B. 0.2–0.25 and 1.65
    C. 1.1 and 0.4–0.5
    D. 0.4–0.5 and 1.1

Answer: A—The electron density (or CET) of a compact bone (e.g., mandible) relative to that of water is taken as 1.65. Spongy bone, such as sternum, has a density of 1.1 g/cm^3. The electron density of lung varies between 0.20 and 0.25 relative to that of water [3, p. 280 (Sects. 14.5.C.1, 14.5.C.2)].

48. What should be the minimum thickness of lead (Pb) shielding to collimate an 18 MeV electron beam?

    A. 5 mm
    B. 10 mm
    C. 18 mm
    D. 19 mm

Answer: B—A rule of thumb may be formulated: The minimum thickness of lead required for blocking in millimeters is given by the electron energy in MeV incident on lead divided by 2. Another millimeter of lead may be added as a safety margin. The required thickness of Cerrobend is approximately 20% greater than that of pure lead [3, p. 287 (Sect. 14.6.B)].

49. Effects of bone on electron dose distribution include:

    A. A small increase in dose upstream due to backscatter (4%) and a small increase in dose inside bone due to multiple Coulomb scattering (7–10%)
    B. A small increase in dose upstream due to backscatter (4%) and a small decrease in dose inside bone due to multiple Coulomb scattering (7–10%)

    C. A small decrease in dose upstream due to backscatter (4%) and a small increase in dose inside bone due to multiple Coulomb scattering (7–10%)

    D. A small decrease in dose upstream due to backscatter (4%) and a small decrease in dose inside bone due to multiple Coulomb scattering (7–10%)

Answer: A—[9, p. 50].

50. A thickness of 3 cm body tissue has to be treated of which 1 cm is bone. What would be the energy required if the treatment depth has to be covered by 90% isodose line?

    A. 4 MeV
    B. 6 MeV
    C. 9 MeV
    D. 12 MeV

Answer: D—Of the total thickness of 3 cm, 2 cm is soft tissue and 1 cm is bone which is equivalent to 1.65 cm of soft tissue. So, total thickness = 2 cm + 1.65 cm = 3.65 cm of soft tissue. Multiply this value by 3.2 (or 3.3) to get the electron energy in MeV. This is approximately 12 MeV, which would cover the required depth by 90% isodose line [5–7].

51. In treating with electron beams, which of the following sites would require internal shielding to protect the normal structures lying beneath them?
(A) Buccal mucosa. (B) Lip. (C) Chest wall.

    A. A and B
    B. B and C
    C. A and C
    D. A, B, and C

Answer: A—In some situations, such as the treatment of lip, buccal mucosa, and eyelid lesions, internal shielding is useful to protect the normal structures beyond the target volume [3, p. 288 (Sect. 14.6.D)].

52. Of the following statements, which one most accurately describes the relationship between electron energy and depth of dose max (D_{max})?

    A. As the energy increases, the D_{max} keeps increasing.
    B. As the energy increases, the D_{max} keeps decreasing.
    C. As the energy increases, the D_{max} decreases initially up to 12 MeV and increases sharply beyond that.
    D. As the energy increases, the D_{max} increases initially up to 12 MeV and decreases gradually beyond that.

Answer: D—Table 7.2 [7].

53. A 9 MeV electron beam is being used. It is desired to protect sensitive structures lying beyond 2 cm depth using internal shielding. What is the lead thickness to be used?

    A. 1 mm
    B. 2 mm
    C. 3.5 mm
    D. 4.5 mm

Answer: C—Incident energy = 9 MeV. Treatment depth = 2 cm. Energy loss = 2 MeV/cm. Energy lost in the first 2 cm = 4 MeV.

Energy at the treatment depth-Pb interface = 9 MeV–4 MeV = 5 MeV.

Lead thickness required = 5 MeV/2 = 2.5 mm of Pb + 1 mm safety margin = 3.5 mm.

54. While treating with electron beams, an internal shielding made of lead is to be used. The resultant electron backscatter is reduced by:

    A. Placing a low-atomic-number absorber between the lead shield and tissue being treated
    B. Placing a low-atomic-number absorber between the lead shield and tissue that is protected by the lead shield
    C. Placing a high-atomic-number absorber between the lead shield and tissue being treated
    D. Placing a high-atomic-number absorber between the lead shield and tissue that is protected by the lead shield

Answer: A—To dissipate the effect of electron backscatter, a suitable thickness of low-atomic-number absorber such as a wax bolus may be placed between the lead shield and the preceding tissue surface [3, p. 289].

55. Consider a treatment site being treated by two *abutting* electron beams—one 9 Mev and one 18 MeV beam. The resultant hotspot will be:

    A. On the side of 18 MeV beam just adjacent to the junction
    B. On the side of 9 MeV beam just adjacent to the junction
    C. On both sides (9 MeV as well as 18 MeV) just below the junction
    D. Negligible in size and can be ignored

Answer: C—Fig. 14.30 on page 284 [3].

56. Consider a treatment site being treated by a photon and an electron beam both *abutting* on the surface. A hotspot will develop on the side of:

    A. The photon field
    B. The electron field
    C. Both photon and electron fields
    D. Either photon or electron field depending on the energies used

Answer: A—When an electron field is abutted at the surface with a photon field, a hot spot develops on the side of the photon field and a cold spot develops on the side of the electron field. This is caused by outscattering of electrons from the electron field [3].

57. The reason for employing one high-energy photon beam and one electron beam combination (coaxial beams, same field size) to treat a single lesion is to:

    A. Deliver more dose to deep-lying structures
    B. Reduce the penumbra in the resultant dose distribution
    C. Widen the penumbra in the resultant dose distribution
    D. Improve the skin-sparing effect and reduce the irradiation of deep underlying tissues

Answer: D—High-energy photons (a few MV or above) improve the skin-sparing effect, while the dose falloff of the electrons reduces the irradiation of deeply underlying tissues [4, p. 64 (Sect. 5.3.1)].

58. As the SSD is extended from the nominal value of 100 cm, the change in output is more with:

    A. High-energy electron beams and large field sizes
    B. High-energy electron beams and small field sizes
    C. Low-energy electron beams and large field sizes
    D. Low-energy electron beams and small field sizes

Answer: D—It can be seen from both sets of data that the air gap correction factors are larger at low electron energies and small field sizes [1, p. 133].

59. Eye shields that are commonly used in electron treatments of eyelids are capable to provide adequate shielding only up to:

    A. 4 MeV or less
    B. 6 MeV or less
    C. 9 MeV or less
    D. 15 MeV or less

Answer: C—[1, p. 135 (Sect. 6.6.1) 10].

60. With electrons, a larger field at the surface may be required to cover PTV adequately than one is usually accustomed to (in the case of photon beams). The reason is:

    A. Steep dose gradient observed in the central axis depth-dose curve
    B. Bulging of the isodose curves less than 50%
    C. Constriction of the 90% and 80% isodose curves
    D. To account for the fact that the practical range of the electrons (in cm) is only just half of the electron energy (in MeV)

Answer: C—Examination of the electron isodose curves reveals that there is a significant tapering of the 80% isodose curve at energies above 7 MeV. The constriction of the useful treatment volume also depends on the field size and is worse for the smaller fields. Thus, with electrons, a larger field at the surface than one is usually accustomed to (in the case of photon beams) may be necessary to cover PTV adequately [3].

References

1. Bova FJ. Clinical electron beam physics. In: Smith AR, editor. Radiation therapy physics. Springer-Verlag: Berlin Heidelberg GmbH; 1995. p. 124–6.
2. Karzmark CJ, Nunan, CS, Tanabe E. Medical electron accelerators. Mc Graw Hill, Inc., 1993. p. 144.
3. Khan FM, Gibbons JP, editors. The physics of radiation therapy. Lippincott Williams & Wilkins; 2014. p. 274. (Section 14.4.H)
4. International Commission on Radiological Units and Measurements (ICRU). Report 71: Prescribing, recording and reporting electron beam therapy, 2004. p. 69.
5. Chang DS et al (eds). Dosimetry of electron beams. In: Basic radiotherapy physics and biology. Springer: New York; 2014. p. 115.
6. Hogstrom KR, Almond PR. Review of electron beam therapy physics. Phys Med Biol. 2006;51(13):R455–89. (Section 6.3)
7. Gerbi BJ, et al. AAPM TG-70 Report: Recommendations for clinical electron beam dosimetry: Supplement to the recommendations of Task Group 25. p. 3243.
8. Podgorsak EB (Tech Ed). Radiation oncology physics: a handbook for teachers and students. International Atomic Energy Agency, Vienna, 2005.
9. Antolak JA, Hogstrom KR. Electron radiotherapy past, present & future. https://amos3.aapm.org/abstracts/pdf/68-19700-237350-85416.pdf.
10. https://civcort.com/ro/miscellaneous/tungsten-eye-shields/tungsten-eye-shields-M19.htm#productOptions

Chapter 6
Proton Therapy

Charles Bloch

(A) Basic Physics

1. Proton therapy lateral penumbra is sharper than photons:

    A. For all treatment depths
    B. For shallow (<17 cm) treatment depths
    C. For deep (>17 cm) treatment depths
    D. For no treatment depths

Answer: B. Per ICRU-78 [1], "for depths typically up to ~17–18 cm in tissue, the lateral dose falloff is steeper than that for photon beams." At larger depths, scattering in the patient increases the lateral penumbra such that it is generally larger than that for photon beams.

2. In a cyclotron, proton energy is increased by:

    A. A large electromagnet
    B. An ion source
    C. RF cavities
    D. Energy selection system (ESS)

Answer: C. All particle accelerators use RF cavities to create an electric field that accelerates charged particles. Electromagnets steer charged particles but do not increase their energy. The ion source produces charged particle. The energy selection system is used to decrease the beam energy via degraders.

C. Bloch (✉)
Radiation Oncology, University of Washington School of Medicine, Seattle, WA, USA
e-mail: cdbloch@uw.edu

M. Heard et al. (eds.), *Absolute Therapeutic Medical Physics Review*,
https://doi.org/10.1007/978-3-031-14671-8_6

3. Continuous beam is provided by a:

   A. Synchrotron
   B. Synchrocyclotron
   C. Cyclotron
   D. None of the above

Answer: C. Isochronous cyclotrons provide a continuous beam. Synchrotrons and synchrocyclotrons provide pulsed beams.

4. A constant magnetic field is used by:

   A. Synchrotron
   B. Cyclotron
   C. A and B
   D. None of the above

Answer: B. Cyclotrons have a large constant magnetic field, and particles of increasing energy move to increasingly large orbits. Synchrotrons are a ring structure and have a fixed orbit requiring the magnetic fields to increase with increasing beam energy to constrain the protons to a fixed orbit.

5. A constant orbit is used by:

   A. Synchrotron
   B. Cyclotron
   C. A and B
   D. None of the above

Answer: A. Synchrotrons have a fixed orbit for the protons, and the magnetic fields ramp up as the proton energy increases to maintain that constant orbit.

6. Compared to scattered proton beams, pencil beam scanning has advantages of:

   A. Lower proximal dose
   B. Lower distal dose
   C. Reduced lateral penumbra
   D. All the above

Answer: A. Passive scattered beams use a compensator to provide distal conformality but pull back the SOBP proximally. Pencil beam scanning (PBS) does not use a fixed SOBP and conform to the target distally and proximally. However, PBS usually forgoes apertures as being unnecessary and tend to have slightly worse penumbra because of that [1].

7. Buildup dose in the entrance region occurs:

    A. In less than 1 mm
    B. Between 1 and 5 mm
    C. Depends on proton beam energy
    D. There is no buildup region for protons

Answer: D. Buildup region in photon therapy refers to buildup of charge equilibrium. Protons are charge particles and will reach equilibrium before interacting with the patient. There is no "buildup" region.

8. Range uncertainty is a result of:

    A. Mis-calibration of CT scanner
    B. Limitations in converting CT numbers to proton stopping powers
    C. Uncertainties in patient composition
    D. All of the above

Answer: D. Proton range is determined by converting CT numbers to proton stopping power; hence, any mis-calibration of the CT scanner will result in a corresponding error in the proton range. Additionally, proton stopping power is not uniquely determined by single-energy CT scans. Materials of different composition can have different proton stopping powers but have identical CT numbers. Additionally, one does not know the actual material composition of any given patient (or implanted materials) and must use "typical" values for patients.

(B) Radiobiology

1. The RBE for proton therapy is defined as the ratio of:

    A. Photon dose to the proton dose required to give the same biological effect
    B. Proton dose to the photon dose required to give the same biological effect
    C. Photon biological effect to proton biological effect for the same dose
    D. Proton biological effect to photon biological effect for the same dose

Answer: A. Per ICRU 78 [1], RBE is the ratio of doses (photon to proton) providing the same biological effect. Because less proton dose is required to provide the same biological effect, the ratio is greater than 1.

2. Proton therapy RBE of 1.1 is due to:

    A. High a/b ratio
    B. Higher beam energies
    C. More DNA double-strand breaks
    D. Neutron production

Answer: C. Higher LET increases the probability of double-strand breaks, which is more likely to lead to cell death.

3. High LET particles produce more:

    A. DNA double-strand breaks
    B. Ionizations within a cell
    C. Ionizations at the end of their range
    D. All of the above

Answer: D. LET increases at the end of range. Higher LET produces more ionizations within the cell, which in turn increases the probability of double-strand breaks.

4. Secondary neutron production is lowest with:

    A. Double scattering
    B. Pencil beam scanning
    C. Uniform scanning
    D. Single scattering

Answer: B. Pencil beam scanning. Single and double scattering produce additional neutrons in the scattering material. Uniform scanning produces additional neutrons in the range compensator and by over-scanning the aperture. Pencil beam scanning eliminates the range compensator, and even if apertures are used, the over-scanning is less.

5. In clinical practice, an RBE of 1.1 is used for:

    A. Proton therapy
    B. Neutron therapy
    C. Carbon therapy
    D. All of the above

Answer: A. Neutron and carbon therapies have high LET, and RBE is much higher than that for protons.

6. Proton RBE increases most rapidly at the end of range resulting in:
    A. An effective range of 1–2 mm greater than the physical range
    B. A sharp rise in cell killing prior to the Bragg peak
    C. Clinical use of a depth-dependent RBE
    D. All of the above

Answer: A. Per ICRU 78 [1], "On the declining distal edge of the SOBP, sharp relative increments (up to 50%) in the RBE have been observed. This results in an effective increase in the range of 1 mm and 2 mm for proton beams in the energy ranges below 75 MeV and above 150 MeV, respectively." This effect is beyond the Bragg peak, not prior to it. While research is investigating the application of a depth-dependent RBE, clinically only a constant value of 1.1 is recommended.

(C) Imaging and Planning

1. In proton therapy, the role of the PTV is:

    A. A tool for determining appropriate beam sizes
    B. Used for prescription and reporting
    C. A and B
    D. None of the above

Answer: C. Per ICRU 78 [1], the PTV has two functions. It can be used to determine beam margins, and it is used by physicians for prescribing and reporting.

2. Ideally, PTV margin should be determined:

    A. By setup uncertainty in all directions
    B. By proton range uncertainty in all directions
    C. Separately for each beam
    D. A and B

Answer: C. Per ICRU-78 [1], setup uncertainty should be used to determine margins perpendicular to the beam direction, while range uncertainty would determine the margin in the beam direction. Since each beam has a different direction, this would result in a unique PTV for each beam.

3. Proton beams may be created based on CTV by:

    A. Using beam-specific margins
    B. Allowing for the fact that the PTV may not be completely covered
    C. Using robust optimization
    D. All the above

Answer: D. Per ICRU-78 [1], beam-specific margins can be used to ensure the coverage of the CTV, which may result in less coverage of the PTV. Robust optimization allows the planning computer to generate the beam-specific margins based on known uncertainties.

4. Proton range uncertainty:

    A. Is the same for all beam energies
    B. Is typically 2–4% of the total range
    C. Must be added on to the PTV margin
    D. All the above

Answer: B. Clinics differ on their estimates of the range uncertainty but typically use numbers around 2–4% of the range. Hence, it varies with beam energy (which determines range). It is not in addition to the PTV margin but rather determines the PTV margin in the beam direction.

5. In proton therapy, "robustness" refers to:

   A. Choosing beam angles to minimize sensitivity to setup variation
   B. Recalculating the dose with simulated setup or range errors
   C. Computer optimization of pencil beams to provide coverage even with setup errors
   D. All of the above

Answer: D. Before computing systems were sufficiently advanced, treatment planning relied heavily on experienced planners to choose "robust" beams, avoiding large heterogeneities parallel to the beam. Faster computer calculations allowed evaluation of range and setup uncertainties to evaluate "robustness," e.g., the ability to maintain coverage under expected accuracy. More advanced planning systems now provide the ability to include robustness as an optimization parameter.

6. Errors in CT calibration can lead to dose errors up to:

   A. 3%
   B. 10%
   C. 50%
   D. 100%

Answer: D. CT calibration determines the range of the protons and hence the position of the high dose. A CT calibration error can shift the high-dose region to an area that was supposed to receive no dose. Alternatively, a high-dose region may fall short of its intended location leaving no dose. Hence, it causes dose errors as large as 100%.

7. Conversion of CT number to stopping power should be done using:

   A. Bethe-Bloch formula
   B. Hounsfield units
   C. Stoichiometric method
   D. Compton scattering

Answer: C. Per ICRU-78 [1] and Schneider et al. [2], the stoichiometric method is recommended for converting CT numbers to proton stopping power, hence water-equivalent density. It relies on a simplified version of the Bethe-Bloch formula. CT number is equivalent to Hounsfield units and depends on photoelectric effect, coherent scattering, and Compton scattering.

8. Because of the significant tissue heterogeneities, proton treatment plans for lung should use:

   A. Pencil beam algorithm
   B. Collapsed-cone convolution algorithm
   C. Monte Carlo algorithm
   D. None of the above

Answer: C. Proton treatment planning systems offer pencil beam algorithms and Monte Carlo algorithms. Collapsed-cone convolution is a photon algorithm. Only the Monte Carlo algorithms offer accurate scatter for heterogeneities.

(D) Prescribing, Reporting, and Plan Evaluation

1. Proton therapy prescriptions should be in:

   A. Gy (RBE)
   B. Cobalt-gray equivalent
   C. Effective dose
   D. Sieverts

Answer: A. Per ICRU-78 [1], Gy is the SI unit of dose. RBE is added to denote the RBE adjustment that has been made. Equivalent dose is reserved for radiation protection which is the unit sievert. Similarly, effective dose is reserved for radiation protection and denotes whole-body dose.

2. Because protons use an RBE of 1.1:

   A. More dose is given
   B. Less dose is given
   C. Local control is 10% higher
   D. None of the above

Answer: B. Per ICRU-78 [1], RBE is the ratio of photon dose to proton dose for the same biological effect. Therefore, an RBE of 1.1 indicates that photon dose is 10% higher than proton dose, or when treating with protons, less physical dose is given.

3. To evaluate a proton plan for robustness, one should look at dose calculations done with:

   A. Typical setup errors (e.g., +/−3 mm)
   B. Typical range errors (e.g., +/−3%)
   C. With and without density overrides
   D. All of the above

Answer: D. Compared to photons, proton dose distributions are more sensitive to setup errors, range errors, and density variations. To be "robust" means to be acceptable given any of these changes.

4. Robust optimization allows the computer to create a dose distribution with a high probability of covering the CTV under expected errors. In these cases, PTV coverage:

   A. May be less than that in conventional planning
   B. Should be at least as high as conventional planning

    C. Should not be reported
    D. A and C

Answer: A. Prescribing PTV coverage provides CTV coverage under all error conditions, not just the most probable. Robust optimization may provide lower coverage of the PTV, but it should still be reported.

(E) Measurement and Verification

1. Absolute calibration of a proton therapy system is done according to:

    A. TG-51
    B. TRS-398
    C. ICRU-78
    D. None of the above

Answer: B. Per ICRU-78 [1], TG-51 does not have a protocol for protons and ICRU-78 is not a calibration protocol.

2. Absolute calibration of a proton beam may be done with:

    A. Cylindrical ion chamber
    B. Parallel plate ion chamber
    C. Calorimeter
    D. All the above

Answer: D, per ICRU-78 [1].

3. The *output* tolerance for *daily* QA is:

    A. 1%
    B. 2%
    C. 3%
    D. Other

Answer: C, per AAPM TG-224 [3]. This is relative to the baseline for the daily QA device.

4. The *daily* QA tolerance for *range* in a pencil beam scanning system is:

    A. 1 mm
    B. 2 mm
    C. 3 mm
    D. Other

Answer: A, per AAPM TG-224 [3]. This is relative to the baseline set for the daily QA device.

5. The *daily* QA tolerance for *spot position* in a pencil beam scanning system is:

    A. 0.5 mm relative, 1 mm absolute
    B. 1 mm relative, 2 mm absolute
    C. 2 mm relative, 3 mm absolute
    D. None of the above

Answer: B, per AAPM TG-224 [3].

6. The *output* tolerance for *monthly* QA is:

    A. 1%
    B. 1.5%
    C. 2%
    D. 3%

Answer: C, per AAPM TG-224 [3]. This is relative to the baseline for the monthly QA device.

7. The *monthly* tolerance for *range* in a pencil beam scanning system is:

    A. 1 mm
    B. 2 mm
    C. 3 mm
    D. Other

Answer: A, per AAPM TG-224 [3]. This is relative to the baseline for the monthly QA device.

8. The *output* tolerance for *annual* QA is:

    A. 1%
    B. 1.5%
    C. 2%
    D. 3%

Answer: C, per AAPM TG-224 [3]. This is based on a TRS-398 calibration (i.e., absolute).

9. The annual tolerance for range in a pencil beam scanning system is:

    A. 1 mm
    B. 2 mm
    C. 3 mm
    D. Other

Answer: A, per AAPM TG-224 [3]. This is for several relevant clinical energies.

10. The annual QA tolerance for spot position for a pencil beam scanning system is:

    A. 1 mm absolute, 0.5 mm relative
    B. 2 mm absolute, 1 mm relative
    C. 3 mm absolute, 2 mm relative
    D. None of the above

Answer: A, per AAPM TG-224 [3].

11. A high-resolution scintillator-based detector is used to measure:

    A. Depth-dose curves
    B. Spot positions
    C. Absolute dose
    D. All of the above

Answer: B, per AAPM TG-224 [3].

12. A multi-leaf ion chamber (MLIC) is used to measure:

    A. Depth-dose curves
    B. Spot sizes
    C. Absolute dose
    D. All the above

Answer: A, per AAPM TG-224 [3].

13. A large-diameter parallel-plate ion chamber is used to measure:

    A. Integrated depth-dose curves
    B. Beam profiles
    C. Absolute dose
    D. All the above

Answer: A, per AAPM TG-185 [4].

References

1. International Commission on Radiation Units and Measurements (ICRU). Prescribing, recording, and reporting proton beam therapy, ICRU Report 78, J ICRU. 2007.
2. Schneider U, Pedroni E, Lomax A. The calibration of CT Hounsfield units for radiotherapy treatment planning. Phys Med Biol. 1996;41(1):111–24. https://doi.org/10.1088/0031-9155/41/1/009.
3. Arjomandy B, Taylor P, Ainsley C, Safai S, Sahoo N, Pankuch M, Farr JB, Yong Park S, Klein E, Flanz J, Yorke ED, Followill D, Kase Y. AAPM task group 224: comprehensive proton therapy machine quality assurance. Med Phys. 2019;46:e678–705. https://doi.org/10.1002/mp.13622.
4. Farr JB, Moyers MF, Allgower CE, Bues M, Hsi W-C, Jin H, Mihailidis DN, Lu H-M, Newhauser WD, Sahoo N, Slopsema R, Yeung D, Zhu XR. Clinical commissioning of intensity-modulated proton therapy systems: Report of AAPM Task Group 185. Med Phys. 2021;48:e1–e30. https://doi.org/10.1002/mp.14546.

Chapter 7
Total Skin Electron Therapy (TSET) and Total Body Irradiation (TBI)

Susha Pillai

1. Total skin electron therapy (TSET) is used for treating what type of cutaneous disease?

    A. Mycosis fungoides
    B. All types of skin cancers
    C. Deep-seated cancers affecting epithelial cells only
    D. Lymphoma cancers

Answer: The correct answer is A. TSET is used for treating a type of chronic lymphoma called T-cell lymphoma, and it is commonly called mycosis fungoides. Lymphocytes are radiosensitive cells and show excellent response with low-level radiation dose [1].

2. What is the approximate size of the useful beam at the treatment distance and the desired beam uniformity in the vertical and horizontal directions?

    A. 40 cm L × 40 cm W; 10% (vertical) and 10% (horizontal)
    B. 200 cm L × 80 cm W; 8% (vertical) and 4% (horizontal)
    C. 100 cm L × 40 cm W; 5% (vertical) and 4% (horizontal)
    D. 300 m L × 100 cm W; 10% (vertical) and 10% (horizontal)

Answer: The correct answer is B. The field size of the composite electron beam at the treatment plane must be at least 200 cm length and 80 cm wide to encompass the entire body. Beam uniformity must be within 8% in the head-to-toe direction and within 4% in the right to left direction. Beam uniformity across the entire treatment volume can be achieved by the selection of largest field size available in high dose

S. Pillai (✉)
Radiation Medicine, Oregon Health and Science University, Portland, OR, USA

© The Author(s), under exclusive license to Springer Nature Switzerland AG 2022
M. Heard et al. (eds.), *Absolute Therapeutic Medical Physics Review*,
https://doi.org/10.1007/978-3-031-14671-8_7

rate (HDR) mode setting in modern linacs, selection of appropriate gantry angles for the dual fields, and use of beam degrader or scatterer [1].

3. What is the most commonly used electron energies for TSET?

    A. Any energies can be used depending on the depth of treatment
    B. < 10 MeV
    C. 6 MeV only
    D. Highest electron energy that is available for the machine

Answer: The correct answer is B. Electron energy of 10 MeV or less can be used for total skin electron therapy due to bremsstrahlung X-ray contamination. Low-energy electron beam is sufficient to deliver the dose to the epidermis and dermis located at a depth of 5 mm [2].

4. What is the most suitable treatment technique for TSET?

    A. Electron arc therapy
    B. Four-field static beam arrangement
    C. Six-dual-field technique
    D. Six-single-field technique

Answer: The correct answer is C. Dual-field technique with rotating platform or six-dual-field technique is the most commonly used treatment technique. While rotational technique provides the most uniform dose distribution over the entire body, the six-dual-field technique is also widely used for TSET. Rotational technique can reduce the treatment setup and treatment time, simplify the beam matching, and compensate for patient motion. Six dual field involves a pair of angled beams with the patient standing in six different orientations spaced at 60° equal intervals, which can also achieve acceptable dose uniformity [1].

5. What is the major role of the beam scatterer in TSET?

    A. Reduce the beam energy
    B. Increase the skin dose
    C. Increase beam uniformity
    D. All of the above

Answer: The correct answer is C. A beam scatterer is placed about 20 cm in front of the patient, and this improves the dose uniformity at the treatment depth particularly on the oblique body surfaces. Wider angular spread of the beam results in a higher skin dose and shallower depth dose due to decreased practical range. An additional scatterer (high-Z materials) can also be placed close to the accelerator exit window [1].

6. Name two major factors that can adversely affect the dose calibration during TSET commissioning:

    A. Treatment distance and machine dose rate setting
    B. Lack of special phantom materials and degrader selection for commissioning
    C. Lack of proper TSET stand and patient-specific shielding materials
    D. Inadequate verification of electron energy at distance, cable effect for the dosimeter

Answer: The correct answer is D. Due to the short range of the electron beam at the treatment distance, care must be taken to choose the thin window chamber for dose calibration. Most commonly used chambers are parallel plate chambers such as the advanced Markus parallel plate chamber. Electric charge created by small-volume chambers is often small, and hence the noise and spurious signals created from irradiating the cable become dominant. Cable-induced effects on ion chamber can be mitigated by shielding the exposed area of the cable using a 5-mm-thick lead shield [1].

7. What is the role of in vivo dosimetry in total skin electron therapy(TSET)?

    A. To verify the uniformity of the dose distribution to the patient's skin
    B. To account for the dosimetric uncertainties in patient positioning during treatment
    C. To identify the regions/areas requiring local boosts
    D. All of the above

Answer: The correct answer is D. The role of in vivo dosimetry in TSET is to verify the uniformity of dose, to account for the dosimetric uncertainties in patient positioning, and to identify regions requiring local boosts [1].

8. Total body irradiation (TBI) is widely used for treating what type of malignancy? What is the most commonly used treatment regime for this malignancy?

    A. Any type of blood cancer; radiation alone
    B. Bone cancer, TBI, and chemotherapy
    C. AML and ALL; combination of chemotherapy, radiation, and bone marrow transplant
    D. All benign and malignant tumors; surgery and radiation

Answer: The correct answer is C. The purpose of the TBI is to destroy the residual cancer cells and suppress the immune response of the recipient and hence reduce the risk of graft rejection. Unlike chemotherapy, radiation allows a uniform dose to the entire body, including the sanctuary sites—brain and testis [3, 4].

9. Which is the most commonly used treatment position for total body irradiation?

   A. Six-dual-field technique
   B. Rotational technique
   C. Translational technique
   D. Standing AP-PA technique and sitting position with opposed lateral technique

Answer: The correct answer is D. AP-PA and lateral techniques are the most commonly used techniques in TBI. Other available treatment positions include sitting position, reclining position, prone-supine (floor technique for pediatric cases), and lateral decubitus. In the recent years, VMAT-based TBI has been gaining interest among clinics [5–7].

10. What is the standard mode of dose prescription, prescription dosage, and fractionation scheme for high-dose TBI regime?

    A. Prescribed to mid-depth at mid-body level, 12–14 Gy in 6–8 fractions BID
    B. Mode of prescription varies from patient to patient; 2 Gy in 1 fraction
    C. Prescribe to anatomic location with maximum separation; no standardized dose/fractionation scheme
    D. Dose prescribed to mid-body location, 4 Gy in 2 fractions; daily treatment

Answer: The correct answer is A. Total doses of 1200 cGy (BID in 8 fractions) for adults and 1320 cGy (BID in 6 fractions) for pediatric cases are the most widely used dose fractionation regimes. Lung blocks made of Cerrobend materials or lead are used to reduce lung toxicity, and the optimal dose rate at the midplane is 8–10 cGy per minute [6, 8].

11. Name the dose-limiting organ in the high-dose TBI regime and standard dose rate/dose constraints used for that organ:

    A. Brain; <10 cGy/min/8–9 Gy max dose
    B. Lungs; <12 cGy/min/8–9 Gy mean lung dose
    C. Spinal cord; <10 cGy/min/12–14 Gy
    D. Kidneys; <10 cGy/min/12 Gy max

Answer: The correct answer is B. Radiation-induced interstitial pneumonitis is the major toxicity in patients receiving total body irradiation. Other acute toxicities include parotitis, nausea, vomiting, dry mouth, mucositis, esophagitis, and alopecia [9, 10].

12. What is the role of degrader and compensator in total body irradiation?

    A. Degrader is used to build up the skin dose; compensator is used for improving the dose homogeneity.
    B. Degrader is used to reduce the dose rate at mid-body; compensator is used to increase the overall skin dose.

   C. Degrader is used to improve the beam flatness; compensator is used to compensate for skin irregularities.

   D. Both degrader and compensator have no known role in improving the dose uniformity.

Answer: The correct answer is A [6].

13. Which of the following statements is true regarding total body irradiation?

   A. Helical tomotherapy and VMAT-based treatment techniques can be employed to achieve better dose homogeneity and dose sparing to critical organs.

   B. A backup treatment plan must be available in case of machine failure.

   C. Performing in vivo dosimetry is a recommended practice for the treatment of total body irradiation.

   D. Influence of TBI on linac workload and use factors must be evaluated and documented for vault shielding.

   E. All of the above.

Answer: The correct answer is E [11–14].

14. Facilities that anticipate the implementation of TBI program must carefully re-evaluate the shielding calculation for the designated vault due to the following reason:

   A. Workload for TBI is much higher due to the extended SSD treatment setup.

   B. Use factor is close to 1 as the radiation is directed at one barrier.

   C. Leakage workload is higher, while the scattered workload is negligible.

   D. All of the above.

Answer: The correct answer is D. Workload can be much higher for TBI due to the extended SSD treatment, and the use factor of 1 must be used as the radiation is always directed at one barrier. It also increases the leakage-radiation contribution to all barriers. Scattered radiation from the isocenter to the secondary barriers is not changed [15].

15. Calculate the TBI workload (Gy/week) if the patient load is 2 cases per week and the prescription dose is 12 Gy in 6 fractions at a treatment distance of 500 cm:

   A. 24 Gy/week

   B. 192 Gy/week

   C. 600 Gy/week

   D. 300 Gy/week

Answer: The correct answer is C [15].

   TBI workload is defined as the weekly total dose at the nominal SSD of 1 m.

   TBI workload = (Total dose (Gy)/week) * (Treatment distance)^2 = 600 Gy/week * 5 * 5 = 600 cGy/week.

References

1. Karzmark CJ. Total skin electron therapy: technique and dosimetry. Int J Radiat Oncol Biol Phys. 1986;12:84–5.
2. Park SY, Ahn BS, Park JM, Ye SJ, Kim IH, Kim JI. Dosimetric comparison of 4 MeV and 6 MeV electron beams for total skin irradiation. Radiat Oncol. 2014;9(1):197.
3. Quast U. Whole body radiotherapy: a TBI-guideline. Med Phys. 2006;31(1):5–12.
4. Wong JYC, Filippi AR, Dabaja BS, Yahalom J, Specht L. Total body irradiation: guidelines from the international lymphoma radiation oncology group (ILROG). Int J Radiat Oncol Biol Phys. 2018;101(3):521–9.
5. Esiashvili N, Lu X, Ulin K, Laurie F, Kessel S, Kalapurakal JA, et al. Higher reported lung dose received during total body irradiation for allogeneic hematopoietic stem cell transplantation in children with acute lymphoblastic leukemia is associated with inferior survival: a report from the Children's oncology group. Int J Radiat Oncol Biol Phys. 2019;104(3):513–21.
6. Van Dyk J, Galvin JM, Glasgow GP, Podgorsak EB. The physical aspects of total and half body photon irradiation. AAPM Report No. 17. 1986;22–24.
7. Springer A, et al. Total body irradiation with volumetric modulated arc therapy: dosimetric data and first clinical experience. Radiat Oncol. 2016;11:46.
8. Wills C, Cherian S, Yousef J, Wang K, Mackley HB. Total body irradiation: a practical review. Appl Radiat Oncol. 2016;5(2):11–7.
9. Lawton CA, Cohen EP, Murray KJ, Derus SW, Casper JT, Drobyski WR, et al. Long-term results of selective renal shielding in patients undergoing total body irradiation in preparation for bone marrow transplantation. Bone Marrow Transplant. 1997;20(12):1069–74.
10. Weshler Z, Breuer R, Or R, Naparstek E, Pfeffer MR, Lowental E, et al. Interstitial pneumonitis after total body irradiation: effect of partial lung shielding. Br J Haematol. 1990;74(1):61–4.
11. Kim DW, Chung WK, Shin DO. Dose response of commercially available optically stimulated luminescent detector, $AI_2O_3{:}C$ for megavoltage photons and electrons. Radiat Prot Dosim. 2012;149(2):101–8.
12. Hui SK, Kapatoes J, Fowler J. Feasibility study of helical tomotherapy for total body or total marrow irradiation. Med Phys. 2005;32(10):3214–24.
13. Ravichandran R, Binukumar JP, Davis CA, Sivakumar SS, Krishnamurthy K, Mandhari ZA, et al. Beam configuration and physical parameters of clinical high energy photon beam for total body irradiation (TBI). Phys Med. 2011;27(3):163–8.
14. Radiation therapy vault shielding calculational methods when IMRT and TBI procedures contribute James E. Rodgers.
15. Structural shielding design and evaluation for megavoltage X- and gamma-ray radiotherapy facilities, NCRP report no. 151.

Chapter 8
Machine QA, Commissioning, and Calibration

Prema Rassiah-Szegedi and Martin Szegedi

1. Gamma index is:

   A. A tool which compares plan quality between two plans.
   B. A measurement of beam energy.
   C. A metric which combines both dose difference and distance to agreement in a single quantity.
   D. Used in conjunction with dose volume histogram to evaluate a plan.

Answer: The correct answer is C. The gamma index is *a QA tool used to compare two sets of dose distribution.* It combines both dose difference (DD) and distance to agreement (DTA) into a single quantity [1, 2].

2. When an IMRT plan with high modulation fails verification QA, what should the physicist do?

   A. Proceed with treatment; the QA is not expected to pass.
   B. Check the TPS commissioning.
   C. Redo IMRT QA.
   D. Consider replanning with an attempt to achieve a less complex intensity pattern.

Answer: The correct answer is D. The planner should consider replanning the IMRT plan and attempt to reduce the complexity of the intensity patterns [3].

P. Rassiah-Szegedi (✉) · M. Szegedi
Radiation Oncology, University of Utah, Huntsman Cancer Hospital,
Salt Lake City, UT, USA
e-mail: prema.rassiah@hci.utah.edu

© The Author(s), under exclusive license to Springer Nature Switzerland AG 2022

M. Heard et al. (eds.), *Absolute Therapeutic Medical Physics Review*,
https://doi.org/10.1007/978-3-031-14671-8_8

3. The purpose of patient-specific IMRT QA is:

   A. To verify the accuracy of the dose delivery system.
   B. To ascertain patient positioning errors.
   C. To check the efficiency of MLC segmentation by the planning system.
   D. To manage patient motion.

Answer: The correct answer is A. Patient-specific QA identifies the discrepancy between the calculated and delivered radiation dose for a specific plan. It continually verifies the performance and accuracy of the delivery system [3].

4. The following should be checked if an IMRT QA fails:

   A. Correct plan version received by record and verify.
   B. Phantom setup.
   C. Accuracy, stability, and calibration of the measurement device.
   D. All of the above.

Answer: The correct answer is D. Incorrect plan version, phantom setup, and the tool used for dose measurement can contribute to the failure of an IMRT plan [3, 4].

5. Which of the following linac QA test is NOT done daily?

   A. Output consistency.
   B. Laser alignment.
   C. Room door interlock.
   D. Flatness and symmetry.

Answer: The correct answer is D. Based on task group reports 40 and 142, output consistency, laser alignment, and door interlock need to be checked daily [5, 6].

6. Which of the following statements is true? Should the QA test tolerances be different for linacs used for SRS versus non-IMRT?

   A. Linacs used for non-IMRT and IMRT have the same MLC QA tests and tolerances.
   B. Linacs used for SRS/SBRT and non-SRS/SBRT have the same imaging QA tests and tolerances.
   C. Daily laser localization tolerances are the same for linacs used for SRS/SBRT and non-SRS/SBRT.
   D. None of the above.

Answer: The correct answer is D. There are more MLC tests for IMRT linacs versus non-IMRT linacs. While the tests, i.e., the imaging tests, are the same for SRS/SBRT and non-SRS/SBRT linacs, some tests have tighter tolerances. Localizing laser and mechanical accuracy is maintained with more stringent requirements with machines delivering SRS/SBRT treatments [5].

7. In 2016, the AAPM released TG 100, the Application of Risk Analysis Methods to Radiation Therapy Quality Management. In this, the AAPM recommends the application of:

    A. Process mapping.
    B. Failure modes (FMs) and effects analysis (FMEA).
    C. Fault tree analysis.
    D. Prospectively address QA.
    E. All of the above.

Answer: The correct answer is E. TG 100 [7] recommends process mapping, FMEA, fault tree analysis, and prospectively addressing QA to manage risk in radiation oncology.

8. The following are the recommended annual MLC QA tests, EXCEPT:

    A. Leaf position repeatability.
    B. MLC spoke shot.
    C. Leaf travel speed.
    D. MLC transmission.

Answer: The correct answer is C. TG 142 recommends leaf travel to be checked monthly [5].

9. The AAPM task group TG 142 recommends that an end-to-end check be done to ensure the fidelity of the overall system whenever a new or revised procedure is introduced. This is:

    A. True.
    B. False.
    C. Partially true, end-to-end test should only be done for new procedures.
    D. Partially true, end-to-end test should only be done for revised procedures.

Answer: The correct answer is A. To ensure fidelity of the overall system, TG 142 recommends an end-to-end check whenever a new or revised procedure is introduced [5].

10. The following are the recommended monthly imaging QA for planar kV imaging, EXCEPT:

    A. Spatial resolution.
    B. Contrast.
    C. Uniformity and noise.
    D. Collison interlock.

Answer: The correct answer is D. It is recommended by TG 142 that collision interlocks be done daily [5].

11. A failed Winston-Lutz test using the kV imaging panel may indicate that.

    A. The imaging isocenter and the radiation isocenter are misaligned.
    B. The flexing of the imaging panel as a function of gantry angle is not corrected.
    C. Positioning the ball bearing used for imaging is incorrect.
    D. All of the above.

Answer: The correct answer is D. All A, B, and C can contribute to a failed Winston-Lutz test [8].

12. Which of the following is false regarding EPID?

    A. It can be used to perform flatness and symmetry.
    B. Its dose-response is linear.
    C. It can be used for dose calibration.
    D. It is dose rate independent.

Answer: The correct answer is C. EPID cannot be used for calibration of dose [4].

13. Which of the following equipment is used during acceptance testing of a linac?

    A. Survey meter.
    B. 3D water tank
    C. Thimble ion chamber.
    D. All of the above.

Answer: The correct answer is D. Survey meters, water tank, and ion chambers are used during acceptance and commissioning [5].

14. The purpose of commissioning a linac is to:

    A. Establish beam characteristics needed for clinical use.
    B. Verify vendor specification of beam qualities.
    C. Check machine constancy.
    D. Verify patient workflow.

Answer: The correct answer is A. The purpose of commissioning is to establish beam data needed for clinical use and for establishing baseline values [5].

15. Among the beam data that is established during commissioning are:

    A. Beam profile.
    B. Percentage depth doses.
    C. Baseline for beam flatness and symmetry.
    D. All of the above.

Answer: The correct answer is D. The purpose of commissioning is to establish beam data needed for clinical use and for establishing baseline values [5].

16. Which of the following is false regarding TG 51 (protocol for clinical reference dosimetry)?

    A. It is used for calibration of photon beams with nominal energies between 60Co and 50 MV.
    B. Calibration can be done using any homogeneous phantom.
    C. It uses ion chambers with dose-to-water calibration factors.
    D. It is used for calibration of electron beams with nominal energies between 4 and 50 MeV.

Answer: The correct answer is B. Based on TG 51, only water should be used as a phantom for dose calibration.

17. Which of the following is true regarding the factors used in the TG 51 formalism?

    A. The quality conversion factor (kQ) is chamber specific.
    B. *Pion* corrects for incomplete ion collection efficiency.
    C. *Ppol* corrects for any polarity effects.
    D. All of the above.

Answer: The correct answer is D [9].

18. What factors do you need to take into consideration when choosing detectors for small field dosimetry data collection?

    A. Volume averaging effect.
    B. Minimal energy, dose, and dose-rate dependence of detector with field size.
    C. A detector with a known correction factor that is preferably close to 1.
    D. All of the above.

Answer: The correct answer is D. Task Group 155 provides recommendations for the measurements of small fields [10].

References

1. Low DA, Dempsey JF. Evaluation of the gamma dose distribution comparison method. Med Phys. 2003;30:2455–64. https://doi.org/10.1118/1.1598711.
2. Pulliam KB, Huang JY, Howell RM, et al. Comparison of 2D and 3D gamma analyses. Med Phys. 2013;41:021710.
3. Miften M, Olch A, Mihailidis D, Moran J, Pawlicki T, Molineu A, Li H, Wijesooriya K, Shi J, Xia P, Papanikolaou N, Low DA. Tolerance limits and methodologies for IMRT measurement-based verification QA: recommendations of AAPM task group no. 218. Med Phys. 2018;45:e53–83. https://doi.org/10.1002/mp.12810.
4. Low DA, Moran JM, Dempsey JF, Dong L, Oldham M. Dosimetry tools and techniques for IMRT. Med Phys. 2011;38:1313–38. https://doi.org/10.1118/1.3514120.
5. Hanley J, Dresser S, Simon W, Flynn R, Klein EE, Letourneau D, Liu C, Yin FF, Arjomandy B, Ma L, Aguirre F, Jones J, Bayouth J, Holme T. AAPM task group 198 report: an implemen-

tation guide for TG 142 quality assurance of medical accelerators. Med Phys. 2021;00:1–56. https://doi.org/10.1002/mp.14992.

6. Kutcher GJ, Coia L, Gillin M, Hanson WF, Leibel S, Morton RJ, Palta JR, Purdy JA, Reinstein LE, Svensson GK, et al. Comprehensive QA for radiation oncology: report of AAPM radiation therapy committee task group 40. Med Phys. 1994;21(4):581–618. https://doi.org/10.1118/1.597316.

7. Huq MS, Fraass BA, Dunscombe PB, Gibbons JP Jr, Ibbott GS, Mundt AJ, Mutic S, Palta JR, Rath F, Thomadsen BR, Williamson JF, Yorke ED. The report of task group 100 of the AAPM: application of risk analysis methods to radiation therapy quality management. Med Phys. 2016;43:4209–62. https://doi.org/10.1118/1.4947547.

8. Benedict SH, Yenice KM, Followill D, Galvin JM, Hinson W, Kavanagh B, Keall P, Lovelock M, Meeks S, Papiez L, Purdie T, Sadagopan R, Schell MC, Salter B, Schlesinger DJ, Shiu AS, Solberg T, Song DY, Stieber V, Timmerman R, Tomé WA, Verellen D, Wang L, Yin FF. Stereotactic body radiation therapy: the report of AAPM task group 101. Med Phys. 2010;37:4078–101. https://doi.org/10.1118/1.3438081.

9. Almond PR, Biggs PJ, Coursey BM, Hanson WF, Huq MS, Nath R, Rogers DW. AAPM's TG-51 protocol for clinical reference dosimetry of high-energy photon and electron beams. Med Phys. 1999;26(9):1847–70. https://doi.org/10.1118/1.598691.

10. Das IJ, Francescon P, Moran JM, et al. Report of AAPM task group 155: megavoltage photon beam dosimetry in small fields and non-equilibrium conditions. Med Phys. 2021;48:e886–921. https://doi.org/10.1002/mp.15030.

Chapter 9
Radiation Protection and Shielding

Max Amurao

1. What is the annual regulatory limit for the total effective dose equivalent for an adult that is occupationally exposed to radiation from radioactive materials?

   A. 1 mSv
   B. 50 mSv
   C. 100 mSv
   D. 150 mSv
   E. 500 mSv

   Answer: The correct answer is B [1]. Note that strictly speaking, this limit is not applicable to radiation-producing equipment, unless it is explicitly referenced by the regulations of a specific state. In general, the "whole body" (head and trunk) limit of radiation-producing equipment is 12.5 mSv per calendar quarter, as referenced in the OSHA regulation [2].

2. What is the dose limit in any 1 h, to an individual member of the public, in any unrestricted area, from external sources of radioactive materials?

   A. 0.02 mSv
   B. 0.05 mSv
   C. 0.10 mSv
   D. 0.15 mSv
   E. 0.50 mSv

   Answer: The correct answer is A [3]. This limit excludes the dose contributions from patients administered radioactive material and released in accordance with 10CFR35.75 [4].

M. Amurao (✉)
Office of Radiation Safety, Washington University in St. Louis, St. Louis, MO, USA
e-mail: maxwell.amurao@wustl.edu

© The Author(s), under exclusive license to Springer Nature Switzerland AG 2022
M. Heard et al. (eds.), *Absolute Therapeutic Medical Physics Review*,
https://doi.org/10.1007/978-3-031-14671-8_9

3. What is the 1-h dose equivalent threshold from radioactive materials for a *high-radiation area*?

   A. 0.02 mSv
   B. 0.05 mSv
   C. 0.1 mSv
   D. 0.5 mSv
   E. 1.0 mSv

   Answer: The correct answer is E [5]. Note that this threshold is measured at 30 cm from the radiation source, or 30 cm from any surface that the radiation penetrates. In addition, this definition is applicable for radioactive material sources. For radiation-producing equipment, a similar threshold exists as described in the OSHA regulations [2] with the caveat that this is for "a major portion of the body." Practical follow-up question: What are the designated high-radiation areas in your clinic?

4. What is the annual regulatory limit for the lens dose equivalent for an adult that is occupationally exposed to radiation from radioactive materials?

   A. 1 mSv
   B. 50 mSv
   C. 100 mSv
   D. 150 mSv
   E. 500 mSv

   Answer: The correct answer is D [1]. Note that strictly speaking, this limit is not applicable to radiation-producing equipment, unless it is explicitly referenced by the regulations of a specific state. In general, the dose to the "lens of the eyes" limit of a radiation-producing equipment is 12.5 mSv per calendar quarter, as referenced in the OSHA regulation [2]. If summed over four quarters, this results in an annual limit of 50 mSv, not 150 mSv as cited by the US NRC.

5. What percentage of the annual regulatory limits for adults is applicable to minors who are occupationally exposed to radiation?

   A. 1%
   B. 5%
   C. 10%
   D. 20%
   E. 50%

   Answer: The correct answer is C. Minors are allowed to work with, and be exposed to radiation from, radioactive materials, but the annual thresholds are significantly lower [6]. This limit is also applicable for exposure from radiation-producing equipment, as referenced in the OSHA regulation [2].

6. What is the average effective dose attributed to natural background for an individual in the USA?

   A. 3.1 mSv
   B. 3.6 mSv
   C. 5.0 mSv
   D. 5.4 mSv
   E. 6.2 mSv

   Answer: The correct answer is A [7, 8].

7. What is the annual regulatory limit for the shallow dose equivalent to the skin of the whole body for an adult that is occupationally exposed to radiation from radioactive materials?

   A. 1 mSv
   B. 50 mSv
   C. 100 mSv
   D. 150 mSv
   E. 500 mSv

   Answer: The correct answer is E. The regulatory annual shallow dose equivalent limit to the skin of any extremity is similar to the "organ dose limit." This limit is similar to the limit for skin of any extremity [1]. Note that strictly speaking, this limit is not applicable to radiation-producing equipment, unless it is explicitly referenced by the regulations of a specific state. In general, the dose limit to the "skin of the whole body" of a radiation-producing equipment is 75 mSv per calendar quarter, as referenced in the OSHA regulation [2]. If summed over four quarters, this results in an annual limit of 300 mSv, not 500 mSv as cited by the US NRC.

8. Radioactive materials that are commonly delivered to radiation oncology (e.g., I-125 eye plaque seeds, Ir-192 HDR source) or nuclear medicine departments (e.g., Tc-99 m MDP, F-18 FDG) should be shipped using what kind of package?

   A. Type A
   B. Type B
   C. Type II
   D. Type III
   E. Type 7

   Answer: The correct answer is A [9]. More detailed description of a Type A package is provided in the US Department of Transportation regulations [10]; relevant values are referenced in the NCRP Report No. 184 [8].

9. What is the annual regulatory limit for the sum of the total effective dose equivalent to an individual *member of the public* from the use of radioactive materials?

   A. 1 mSv
   B. 50 mSv
   C. 100 mSv
   D. 150 mSv
   E. 500 mSv

   Answer: The correct answer is A [3]. This limit excludes the dose contributions from background radiation, from any administration the individual has received, from exposure to individuals administered radioactive material and released under § 35.75, from voluntary participation in medical research programs, and from the licensee's disposal of radioactive material into sanitary sewerage in accordance with § 20.2003.

10. What is the regulatory limit for the dose equivalent to the embryo/fetus during the entire pregnancy, due to the occupational exposure of a declared pregnant woman?

    A. 0.5 mSv
    B. 1 mSv
    C. 5 mSv
    D. 10 mSv
    E. 50 mSv

    Answer: The correct answer is C [11]. Note that this threshold is hinged on the written declaration of pregnancy of the radiation worker. If the dose equivalent to the embryo/fetus exceeds 4.5 mSv by the time the woman declares the pregnancy, the additional dose equivalent to the embryo/fetus should not exceed 0.5 mSv during the remainder of the pregnancy.

    For pregnant patients undergoing photon radiation therapy, AAPM TG 36 talks about the effects of radiation on developing fetus and the professional considerations to minimize the dose [12].

11. What is the 1-h dose equivalent threshold from radioactive materials for a designated *radiation area*?

    A. 0.02 mSv
    B. 0.05 mSv
    C. 0.1 mSv
    D. 0.5 mSv
    E. 1.0 mSv

    Answer: The correct answer is B [5]. Note that this threshold is measured at 30 cm from the radiation source, or 30 cm from any surface that the radiation penetrates. In addition, this definition is applicable for radioactive material sources. For radiation-producing equipment, a similar threshold exists as described in the OSHA regulations [2] with the caveat that this is for "a major portion of the body," with an additional caveat of a 1 mSv threshold for any 5 consecutive days. Practical follow-up question: What are the designated radiation areas in your clinic?

12. Which of the radiopharmaceuticals listed below *does not* require a written directive, signed and dated by the authorized user prior to administration to a patient?

   A. Ra-223 Xofigo, 100 microcuries
   B. F-18 FDG, 15 millicuries
   C. I-131 sodium iodide, 2 millicuries
   D. Y-90 TheraSphere, 5 GBq (135 millicuries)
   E. Lu-177 Lutathera, 200 millicuries

   Answer: The correct answer is B. 15 millicuries of F-18 FDG is typically used for PET/CT oncologic imaging. It is relevant to note that the correct answer is not solely based on the magnitude of the activity administered. The emission type is a relevant consideration, where alpha-emitters and beta-emitters usually require a written directive in typical clinical use.

13. What is the annual regulatory limit for the sum of the deep-dose equivalent and the committed-dose equivalent to any individual organ or tissue other than the lens of the eye for an adult that is occupationally exposed to radiation?

   A. 1 mSv
   B. 50 mSv
   C. 100 mSv
   D. 150 mSv
   E. 500 mSv

   Answer: The correct answer is E. The annual occupational dose limit for each organ is 500 mSv [1]. Note that strictly speaking, this limit is not applicable to radiation-producing equipment, unless it is explicitly referenced by the regulations of a specific state. In general, the dose limit to active blood-forming organs or the gonads from radiation-producing equipment is 12.5 mSv per calendar quarter, as referenced in the OSHA regulation [2]. If summed over four quarters, this results in an annual limit of 100 mSv, not 500 mSv as cited by the US NRC.

14. Which of the following is *not* a standard procedure for receiving and opening radioactive White I, Yellow II, or Yellow III packages?

   A. Monitor the package within 3 h of receipt if delivered during normal working hours.
   B. Monitor the radiation levels at 1 m away from the external surface of the package.
   C. Monitor the external surface of the package for removable radioactive contamination.
   D. Monitor the external surface of the package for degradation of package integrity.
   E. Monitor the access of the delivery courier to the hot lab once every quarter.

Answer: The correct answer is E. The required procedures for receiving and opening packages containing radioactive materials are described in [13]. Additional details on model procedures for safely opening packages containing radioactive materials for medical use are provided in the US NRC Technical Report NUREG 1556 [14].

15. A state that has signed an agreement with the US Nuclear Regulatory Commission receiving authorization to regulate certain uses of radioactive materials within the state is called a/an [15]:

    A. NRC-approved state
    B. NRC-controlled state
    C. NRC-agreement state
    D. NRC-dependent state
    E. NRC-supervised state

    Answer: The correct answer is C. The US NRC relinquishes to the (agreement) state portions of its regulatory authority to license and regulate by-product materials (radioisotopes), source materials (uranium and thorium), and certain quantities of special nuclear materials.

16. What is the largest contributor to an average individual's effective dose from "man-made" radiation sources in the USA?

    A. Computed tomography scans
    B. Nuclear medicine and PET imaging
    C. Cosmic, solar, and satellite radiation
    D. Interventional radiology procedures
    E. 5G mobile broadband communication

    Answer: The correct answer is A. The relevant values are referenced in the NCRP Report No. 184 [8].

17. A radioactive material package was delivered to the clinic, with a measured maximum surface exposure rate of 15 mR/h and a maximum exposure rate of 1.8 mR/h at 1 m away. What should the package be labeled as?

    A. Yellow III
    B. Yellow II
    C. White III
    D. White II
    E. White I

    Answer: The correct answer is A. A Yellow II package has a maximum surface exposure rate reading between 0.5 mR/h and 50 mR/h (satisfied by this package), as well as a maximum surface exposure rate reading at 1 m away between 0.05 mR/h and 1.0 mR/h (not satisfied by this package).

18. Which of the following is *not* required when calculating the barrier protection factor (B)?

    A. Workload
    B. Occupancy

C. Shielding material
D. Dose limit (at shielding point)
E. Distance (source-to-shielding point)

Answer: The correct answer is C. Accounting for the shielding material is relevant after calculating the barrier protection factor.

19. What is the weekly permissible dose limit for a controlled area where a radiation worker may be exposed?

    A. 0.02 mSv/week
    B. 0.05 mSv/week
    C. 0.1 mSv/week
    D. 1 mSv/week
    E. 5 mSv/week

Answer: The correct answer is C. This stems from the regulatory limit of 5.0 mSv for the total effective dose equivalent for a minor radiation worker. It is also relevant for the regulatory limit to the embryo/fetus of a declared pregnant woman spanning the entire course of pregnancy. Although pregnancy covers a 40-week period, assuming an effective cumulative exposure of 50 weeks provides an additional safety factor.

20. When calculating the *primary barrier* protection factor for an HDR brachytherapy suite, what is the appropriate use factor (U)?

    A. 1
    B. 1/2
    C. 1/5
    D. 1/8
    E. 1/40

Answer: The correct answer is A. All barriers of an HDR brachytherapy suite are for the "primary beam."

21. What is the typical occupancy factor (T) assigned to an adjacent clinical area such as a patient exam room, procedure or imaging suite, or a linear accelerator vault?

    A. 1
    B. 1/2
    C. 1/5
    D. 1/8
    E. 1/20

Answer: The correct answer is B. This is described in the NCRP 147 and NCRP 151 [16, 17].

22. For a linear accelerator vault designed to be operated such that neutron production is not an issue, the total barrier protection factor (B_{tot}) can be calculated from the primary (B_{pri}), secondary (B_{sec}), and leakage (B_{leak}) barrier protection factors using ______. Note that HVL is half value layer and TVL is tenth value layer:

A. $B_{tot} = B_{pri} + B_{sec} + B_{leak}$
B. $B_{tot} = B_{pri} \times B_{sec} \times B_{leak}$
C. $B_{tot} = B_{pri} + B_{sec} + B_{leak} + 1\text{HVL}$
D. $B_{tot} = B_{pri} \times B_{sec} \times B_{leak} \times 1\text{HVL}$
E. $B_{tot} = B_{pri} + B_{sec} + B_{leak} + 1\text{TVL}$

Answer: The correct answer is B. The total barrier protection factor is a product of the primary, secondary, and leakage barrier protection factors.

23. The "protection point" located on the same floor/level (neither above nor below) is taken at what distance behind the protective barrier?

A. 0.1 m
B. 0.3 m
C. 0.5 m
D. 1.0 m
E. 3.0 m

Answer: The correct answer is B. This is described in the NCRP 147 and NCRP 151 [16, 17].

24. When considering secondary and leakage shielding, what additional shielding needs to be added to the thicker barrier of the two (secondary and leakage) if these are within one tenth value layer (TVL) of each other?

A. None.
B. 1 HVL
C. 3 HVLs
D. 1 TVL
E. 3 TVLs

Answer: The correct answer is B. As a rule of thumb, when the secondary and leakage shielding barriers are calculated to be within one TVL, one additional HVL is expected for the final barrier to be installed.

25. What is the weekly permissible dose limit for an uncontrolled area where a member of the public may be exposed?

A. 0.02 mSv/week
B. 0.05 mSv/week
C. 0.1 mSv/week
D. 1 mSv/week
E. 5 mSv/week

Answer: The correct answer is A. This stems from the annual regulatory limit of 1.0 mSv for a member of the public, assuming an effective cumulative exposure of 50 weeks.

26. Which of the following clinical modalities listed below will need consideration for the use of borated polyethylene (BPE) as a barrier shielding material?

   A. Linac, 15X
   B. Cyberknife
   C. Gamma Knife
   D. Linac, 12 MeV
   E. HDR brachytherapy

Answer: The correct answer is A. BPE is an effective material for shielding neutrons which are produced when photons exceeding 10 MV are used. A Cyberknife emits 6 MV photons, a Gamma Knife emits 1.17 MeV and 1.33 MeV Co-60 photons, a 12-MeV linac produces electrons, and an HDR unit emits a variety of gamma photons, beta particles, and electrons, but not one of these treatment modalities provide the suitable conditions for neutron production.

27. Which of the following is *not* required when calculating the primary barrier protection factor (B_{pri}) for a dedicated room where radioactive materials are used?

   A. Use factor
   B. Workload
   C. Radioisotope
   D. Exposure rate constant
   E. Maximum (radio)activity

Answer: The correct answer is A. The use factor is one (1) for a dedicated room where radioactive materials are used, such as a Gamma Knife vault or an HDR brachytherapy vault.

28. What is the typical occupancy factor (T) assigned to the point behind the door of a dedicated room where radiation is produced or emitted?

   A. 1
   B. 1/2
   C. 1/5
   D. 1/8
   E. 1/20

Answer: The correct answer is D. The location behind the door leading to a room that is dedicated for the use of radiation (e.g., linac vault, CT simulator suite scanner, HDR vault) is assigned an occupancy factor of 1/8, as described in both the NCRP 147 and NCRP 151. Note that surgical suites and procedure rooms such as those used for LDR brachytherapy are not dedicated radiation-use areas.

29. When calculating the *secondary barrier* protection factor for a CT scanner used either as a CT simulator or as a diagnostic CT scanner, what is the appropriate use factor (U)?

    A. 1
    B. 1/2
    C. 1/5
    D. 1/8
    E. 1/40

    Answer: The correct answer is A. The detectors of the CT scanner serve as the primary barrier. All barriers of a CT scanner are for the scattered and leakage photons.

30. A caregiver present *in* the CT-sim room during image acquisition of an "eyes-to-thighs" scan will receive the least radiation dose by standing at ____:

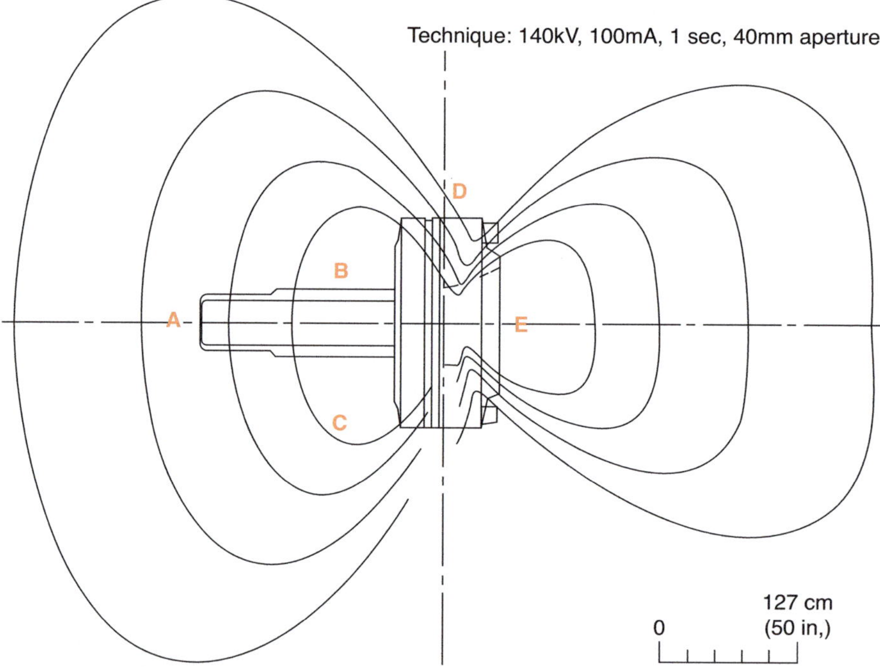

    A. Point A, the foot of the patient scanner bed
    B. Point B, to the side of the patient scanner bed
    C. Point C, three feet away from the side of the patient scanner bed
    D. Point D, next to the CT scanner gantry
    E. Point E, near the head-end of the scanner bore

    Answer: The correct answer is D. The "self-shielding" provided by the detectors that serve as the primary barrier, as well as other CT scanner components mounted on the gantry, significantly reduces the radiation dose adjacent to the scanner gantry.

31. What is a typical minimum barrier (thickness and material) of a CT-sim suite wall?

    A. 1.6 mm lead equivalent
    B. 3.2 mm lead equivalent
    C. 10 mm lead equivalent
    D. 20 mm standard-density concrete
    E. 40 mm low-density concrete

Answer: The correct answer is A. The typical barrier of a CT-sim suite wall is at least 1/16th inches of lead equivalent material, which is approximately 1.6 mm of lead-equivalent material.

An HDR vault is being planned in a single-story outpatient facility, with a very simple floorplan shown below. The RAM License allows up to 15 curies of Ir-192. Other relevant factors and constants are as follows:

$\Gamma = 4.6$ R cm^2/mCi hr.	f-factor = 0.964 cGy/R
$T_{outdoor} = 1/40$	$T_{waiting} = 1/20$
$T_{control} = 1$	$T_{CT\text{-}Sim} = 1/2$

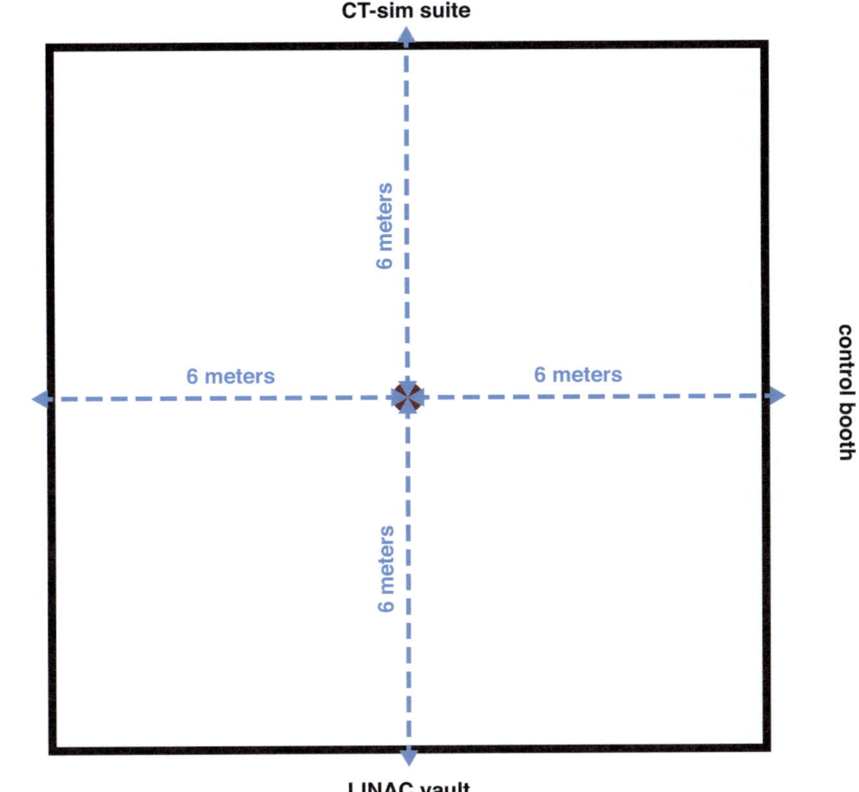

32. Calculate the weekly dose at a protection point 6 m away, assuming that the newly installed 15 curie source is unshielded at the center of the vault for a total of 20 h per week, inclusive of patient treatment time and quality control. For simplicity, assume that the activity remains constant for the week being considered:

   A. 37 mGy/week
   B. 22 Gy/week
   C. 16 mGy/week
   D. 8 mGy/week
   E. 4 mGy/week

Answer: The correct answer is A. The workload can be calculated from:

$$D_{week} = A \times \Gamma \times t \times f \div r^2$$

$$D_{week} = (15{,}000\,\text{mCi}) \times (4.62\,\text{R·cm}^2/\text{mCi·h}) \times (20\,\text{h/week}) \times (9.64\,\text{mGy/R}) \div (600\,\text{cm})^2$$

$$D_{week} = 37\,\text{mGy/week}$$

33. What is the barrier protection factor for the control booth?

   A. 0.263
   B. 0.125
   C. 0.062
   D. 0.045
   E. 0.027

Answer: The correct answer is E. $B = P \cdot d^2 / W \cdot U \cdot T = 1\,\text{mGy/week.} \div 37\,\text{mGy/week} = 0.027$.

34. How many TVLs of standard-density concrete would be needed for the wall to the control booth?

   A. 1.6
   B. 1.3
   C. 1.2
   D. 0.9
   E. 0.6

Answer: The correct answer is A. $n = -\log(B) = -\log(0.027)$.

35. What is the barrier protection factor for the outdoor parking lot?

   A. 1.8×10^{-2}
   B. 2.2×10^{-2}
   C. 2.7×10^{-3}
   D. 6.2×10^{-3}
   E. 5.4×10^{-4}

Answer: The correct answer is B. $B = P \cdot d^2 / W \cdot U \cdot T = 0.02\,\text{mGy/week} \div (37\,\text{mGy/week} \times 1/40) = 0.022$.

36. How many TVLs of high-density concrete would be needed for the exterior wall to the parking lot?

    A. 1.2
    B. 1.7
    C. 2.2
    D. 2.6
    E. 3.3

    Answer: The correct answer is B. $n = -\log(B) = -\log(0.022)$.

37. According to TG 203, which of the following statement is correct?

    A. Use of a lead shield would reduce the dose to the pacemaker for a VMAT plan whose field border is 10 cm inferior to the implantable device.
    B. A lead shield is recommended to reduce the internally scattered radiation.
    C. A lead shield would completely attenuate the photons leaking from the accelerator head.
    D. None of the above statements are true.

    Answer: The correct answer is D. Task Group 203 does not recommend the use of a lead shield for treatment. Treatment planning should utilize beam angles to increase the distance between the field device [18].

References

1. https://www.nrc.gov/reading-rm/doc-collections/cfr/part020/part020-1201.html (10CFR20.12019(a)(1)(i).
2. https://www.osha.gov/laws-regs/regulations/standardnumber/1910/1910.1096 29CFR1910.1096(b)(1).
3. https://www.nrc.gov/reading-rm/doc-collections/cfr/part020/part020-1301.html 10CFR20.1301(a)(2).
4. https://www.nrc.gov/reading-rm/doc-collections/cfr/part035/index.html 10CFR35.75.
5. https://www.nrc.gov/reading-rm/doc-collections/cfr/part020/part020-1003.html 10CFR20.1003
6. https://www.nrc.gov/reading-rm/doc-collections/cfr/part020/part020-1207.html 10CFR20.1207
7. Kase KR, Strom DJ, Birchall A, Brenner DJ, O'Brien III K, Borak TB, Goldhagen PE, Puskin JS. NCRP report no. 160, ionizing radiation exposure of the population of the United States.
8. Mettler FA, Mahesh M, Chatfield MB, Chambers CE, Elee JG, Frush DP. NCRP Report No. 184–Medical radiation exposure of patients in the United States 2019.
9. https://www.nrc.gov/reading-rm/doc-collections/cfr/part071/index.html 10CFR71.4
10. https://www.law.cornell.edu/cfr/text/49/173.415 49CFR173.415.
11. https://www.nrc.gov/reading-rm/doc-collections/cfr/part020/part020-1208.html 10CFR20.1208(a).
12. AAPM. Report no 50 fetal dose from radiotherapy with photon beams, task group 36. Med Phys. 1995;22(1):63–82.

13. https://www.nrc.gov/reading-rm/doc-collections/cfr/part020/part020-1906.html 10CFR20.1906.
14. US NRC Technical Report NUREG 1556 Volume 9, Revision 3, Appendix P published in 2019.
15. https://www.nrc.gov/reading-rm/basic-ref/glossary/agreement-state.html
16. Brunette JJ. NCRP report no. 147, structural shielding design for medical X-ray imaging facilities. Health Phys. 2005;89(2):183.
17. Glasglow GP. Report no. 151–structural shielding design and evaluation for megavoltage X- and gamma-ray radiotherapy facilities. 2005;33(9):3578.
18. Miften M, et al. Report no.203–Management of radiotherapy patients with implanted cardiac pacemakers and defibrillators: a report of the AAPM TG-203. Med Phys. 2019;46(12):e757–88.

Chapter 10
Incidents and Reporting

Erli Chen and Benjamin B. Williams

1. What medical events are commonly seen in radiation oncology?

    A. Wrong patient or treating incorrect site.
    B. Delivered dose different than prescribed dose.
    C. Patient fall.
    D. All of the above.

Answer: Correct answer is D. The most common medical events that happen in radiation oncology departments [1, 2] are treatment of patient A with patient B's plan, treatment of the incorrect site due to incorrect setup from a prior treatment's tattoo marks, missing performance of plan or initial chart check prior to start of treatment to verify the prescription or execute a modified prescribed dose in time, and patient fall due to fatigue through a course of treatment. This is why the NRC (United States Nuclear Regular Commission), state, and the Joint Commission regulations require performance of a time-out procedure prior to each radiation treatment. Performance of initial chart checks, weekly chart checks, daily output checks, and machine monthly QA reduces this risk. Treatment fatigue is the most common side effect for radiation therapy, and patient fall risk needs to be evaluated regularly throughout the course of the treatment. See reference [3] NRC 10 CFR 35 for details.

E. Chen (✉)
Radiation Oncology Department, Cheshire Medical Center, Dartmouth Cancer Center Keene, Keene, NH, USA
e-mail: echen@cheshire-med.com

B. B. Williams
Radiation Oncology Department, Dartmouth Hitchcock Medical Center, Dartmouth Cancer Center, Lebanon, NH, USA
e-mail: Benjamin.B.Williams@hitchcock.org

© The Author(s), under exclusive license to Springer Nature Switzerland AG 2022
M. Heard et al. (eds.), *Absolute Therapeutic Medical Physics Review*,
https://doi.org/10.1007/978-3-031-14671-8_10

2. What are reportable medical events or misadministrations as defined by the NRC 10 CFR 35.3045 or via agreement state regulations?

   A. Wrong patient.
   B. Treatment of incorrect site or dose delivered by the wrong mode of treatment.
   C. Administration of wrong isotope or wrong route of administration.
   D. Delivered dose different by more than 20% of total prescribed dose or 50% of a single fraction.
   E. All of the above.

Answer: Correct answer is E. Review NRC 10 CFR 35.3045 and your agreement state radiation activity material (RAM) license regulation definitions. Agreement state regulation may differ from the NRC's. See reference [3] NRC 10 CFR 35 for details.

3. What are the incident learning systems commonly used in radiation oncology?

   A. Chart rounds.
   B. Tumor board.
   C. M&M conference (morbidity and mortality conference).
   D. Peer review of GTV, CTV, PTV, and OAR contours.
   E. ROILS (Radiation Oncology Incident Learning System) [1].
   F. All of the above.

Answer: Correct answer is F. In radiation oncology, chart round, tumor board, and peer review of GTV, CCTV, PTV, and OAR contours have been used to prevent errors, reduce near misses, and learn about the program and system deficiencies. ROILS and M&M conferences are used to record, analyze, and share findings to improve patient safety and reduce future system risk of medical events.

4. During pretreatment initial plan and chart reviews, you have discovered that a patient was consented for left-breast treatment, but the radiation oncologist prescribed for right-breast treatment. What should you do?

   A. It is not a big deal; side effects are the same for either breast site.
   B. I have never checked a patient's consent; it is not my job.
   C. Inform and discuss with the prescribing radiation oncologist to ensure that the correct site has been consented and prescribed, and then complete the initial plan and chart review, including a patient alert to correct consent prior to first treatment, if needed.
   D. Stop checking plan and chart until the patient has been re-consented.

Answer: Correct answer is C. Based on the MPPG 11.a [4] and TG 100 [5] RPN evaluation, since this event has been discovered prior to treatment, the severity of this event to cause patient harm or treatment dose variation is very low. The best action plan to provide continuous patient care is to inform and discuss with the prescribing radiation oncologist to ensure that the correct site has been consented and prescribed, and then complete the initial plan and chart review, including a patient alert to correct consent prior to first treatment if needed. If no verification has been done, the patient may be treated on the wrong-side breast, which is a severe medical event requiring a phone notification to the NRC or state regulator no later than the next business day after discovery of the event and submission of a written report within 15 days after discovery of the event.

5. During pretreatment initial plan and chart review, you have discovered that a patient has a large gas bubble in the rectum during simulation. The physician prescribed a prostate IGRT-IMRT/VMAT plan to 7920 cGy for this patient. What should you do?

   A. Since the patient will be imaged with CBCT daily and treated with VMAT plan, this plan is clinically acceptable.
   B. Inform the radiation oncologist that it is very difficult to reproduce same rectum volume and PTV planning conditions during daily treatment. Relative prostate and rectum position uncertainty will exceed normal PTV margin limits and rectum may receive much higher dose, and then plan DVH shown.
   C. Stop checking plan and chart until the patient has been re-simulated.
   D. It is not my place to say anything since the physician has reviewed and approved this plan.

Answer: Correct answer is B. Based on the MPPG 11.a standard [4], a plan review should include treatment plan images, OAR contours, isodose coverage, and DVH reviews. You should inform the radiation oncologist that it is very difficult to reproduce the same rectum volume and PTV planning conditions during daily treatment. Relative prostate and rectum position uncertainty will exceed normal PTV margin limits and rectum may receive much higher dose, and then plan DVH shown. IGRT with daily CBCT could reduce some of the position uncertainty; however, the best practice [6] is to provide pre-simulation patient education on how to prepare bowels for upcoming simulation. After your discussion with the radiation oncologist, document and inform therapists of any patient pretreatment instruction or setup changes needed.

6. During Y^{90} microsphere pretreatment activity verification, you have discovered that the ordered activity is 10% different than the prescribed activity. What should you do?

   A. Do nothing; it is within acceptable limit.
   B. I do not know; we do not provide Y^{90} microsphere treatment.
   C. Inform the ordering physician that the difference will increase the possibility of a medical event/misadministration.
   D. Perform a new calculation and then recommend a new treatment time that may better match the prescribed activity.
   E. C + D.

Answer: Correct answer is E. Although a 10% difference between the ordered and prescribed activity is within acceptable limits, it is on the edge of the limit. Starting with this large difference increases the possibility of a medical event/misadministration. The best practice is to perform a new calculation, have a discussion with the ordering physician, and recommend a new treatment time that may better match the prescribed activity to reduce the possibility of a medical event/misadministration.

7. Your hospital is planning to start a new lung SBRT program. How should you evaluate program readiness?

    A. Use TG 100 [5] to map the process, analyze risks, and develop a checklist based on RPN to make sure that the riskiest and most severely consequential steps in the process can be assessed for QA and evaluated.
    B. Attend vendor training.
    C. Have an expert provide peer review for the first three cases.
    D. All of the above.

Answer: Correct answer is D. You should use TG 100 [5] to evaluate any new programs. Map the processes, and analyze and calculate risks based on severity, occurrence, and detectability. Develop a checklist based on RPN to make sure that the riskiest and most severely consequential steps in the process can be assessed for QA and evaluated. Provide step-by-step procedures and special training and have an expert provide peer review for the first three cases.

8. Based on TG 100 and ROILS recommendations, what are the elements you should report to support root cause analysis?

    A. Definition of the event.
    B. Causality.
    C. Severity.
    D. Process maps.
    E. Data elements.
    F. All of the above.

Answer: Correct answer is F. To better support root cause analysis, as well as learning and sharing findings, TG 100 and ROILS [2] have recommended that all events should report five base elements, i.e., definition of the event, causality, severity, process maps, and data elements.

9. Which statement is correct? ASTRO Radiation Oncology Incident Learning System (ROILS) [2] has identified that.

    A. The most common contributing factor for all errors is miscommunication.
    B. Treatment planning is the most commonly identified workflow step where an event occurs.
    C. Pretreatment QA, treatment delivery, and on-treatment imaging are the most commonly identified workflow steps where events are discovered.

D. The majority of commonly identified dose deviations for radiation treatment incidents that did not affect multiple patients are under 5% variation from total prescribed dose.

E. All of the above.

Answer: Correct answer is E. Per ROILS studies, the most common contributing factor for all errors is miscommunication. Treatment planning is the most commonly identified workflow step where an event occurs. Pretreatment QA, treatment delivery, and on-treatment imaging are the most commonly identified workflow steps where events are discovered. The majority of commonly identified dose deviations for radiation treatment incidents that did not affect multiple patients are under 5% variation from total prescribed dose.

10. Per ROILS [1, 2] studies, what percentage of near-miss events have been discovered and can be prevented through the pretreatment QA process?

    A. 50%
    B. 60%
    C. 70%
    D. I have not read any ROILS reports.

Answer: Correct answer is B. Per ROILS reports, 60% of all events are discovered prior to a patient's first treatment delivery. These events are discovered 5% before simulation, 10% during preplanning imaging and simulation, 20% during treatment planning process, 25% by pretreatment QA review such as initial plan and chart review, 25% during treatment delivery, 5% by on-treatment QA (weekly chart check and port film), 2% after treatment course is completed, 5% outside the radiation therapy workflow, and 2% related to equipment and software issue.

11. During an SRS treatment physics final chart check, you have discovered that a patient has two SRS plans, with the first plan designed to treat metastases 1 through 5 and a second plan designed to treat metastases 6 through 11. Both plans have been reviewed and approved by the radiation oncologist. The first plan has been reviewed, IMRT QA was performed and analyzed by a physicist, but these were not done for the second plan. Both plans were scheduled for treatment and are 1 week apart. For the first treatment, the patient was treated with the second plan prior to physicist review and IMRT QA. How should you report this event?

    A. Notify the state or NRC within 24 h, because the wrong site was treated.
    B. No report is needed; the patient needs treatment for all metastases with delivery of both plans.
    C. Inform the attending radiation oncologist and department manager and complete an incident report.
    D. Complete the second SRS plan QA ASAP, perform a root cause analysis, and provide a possible process improvement plan.
    E. Since it is a single treatment, we do not do physics end-of-treatment chart check for SRS or SBRT cases.

Answer: Correct answer is C. Per MPPG 9.a and MPPG 11.a, AAMP, ACR, and ASTRO quality and safety standards [6, 7] call for performance of an end-of-treatment final physics chart check. In this case, since there is no dose deviation from the prescribed dose, the level and severity of this event should be evaluated and determined by the department manager and then reported according to your hospital, departmental policy, and RAM license requirements. It is the best practice to complete the second SRS plan QA ASAP, perform a root cause analysis, and provide a possible process improvement plan.

12. All statements below are correct, except which one?
    I can share radiation-related medical event reports with

    A. NRC or state RAM license regulatory staff.
    B. FDA and the Joint Commission.
    C. In-house risk management and radiation safety committee.
    D. ROILS, when the hospital legally approved enrollment in this program.
    E. Share or post online anywhere I like without hospital legal approval.

Answer: Correct answer is E. Radiation-related medical event reports should be shared with in-house risk management and the radiation safety committee and could be shared with appropriate regulatory agents such as NRC, state RAM license, FDA, and the Joint Commission. These reports can only be shared with ROILS when your hospital has legally approved enrollment in this program. It is inappropriate to share these reports with any other unapproved outside entities or post them online. There may be legal consequences to sharing these reports without your hospital approval.

Study material and reference: AAPM Medical Physics Practice Guidelines (MPPG 9.a, MPPG 11.a) [4], AAPM Task Group report 100 (TG 100) [5], ASTRO ROILS reports [1, 2], NRC 10 CFR 35 [3], ACR-AAPM radiation oncology practice standards [6], and ASTRO Safety is No Accident 2019 [7].

References

1. ROILS system—https://www.astro.org/Patient-Care-and-Research/Patient-Safety/RO-ILS
2. ROILS report—https://www.astro.org/ASTRO/media/ASTRO/Patient%20Care%20and%20Research/PDFs/ROILS_2021_Q2.pdf
3. NRC 10 CFR 35—Subpart M—Reports I NRC.gov
4. AAPM MPPG reports—https://www.aapm.org/pubs/MPPG/?od1n
5. AAPM TG 100—The report of Task Group 100 of the AAPM: Application of risk analysis methods to radiation therapy quality management-Huq-2016-Medical Physics-Wiley Online Library
6. ACR-AAPM technical standard for the performance of radiation oncology physics for external beam therapy–https://www.acr.org/-/media/ACR/Files/Practice-Parameters/ext-beam-ts.pdf
7. ASTRO SINA report—https://www.astro.org/Patient-Care-and-Research/Patient-Safety/Safety-is-no-Accident

Chapter 11
Integrity, Professionalism, and Ethics

Nina Bahar

Part I: Principles and General Guidelines of TG 109
Unless otherwise specified, select all applicable answers.

1. In TG 109, which of the following is not a principle [1]?

    A. "Members must strive to provide the best quality patient care and ensure the safety, privacy, and confidentiality of patients and research participants."
    B. "Members must strive to be impartial in all professional interactions, and must disclose and formally manage any real, potential, or perceived conflicts of interest."
    C. "Members must act with integrity in all aspects of their work."
    D. 'Members must harmonize the Code of Ethics with other codes of conduct to which they are bound.'
    E. None of the above.

Answer: D. TG 109 does say that members of the AAPM are responsible for "harmonizing the code with other codes to which they are bound" in the Preamble, but it is not a core principle that is interpreted across different contexts in ensuing sections. Choices A–C are quoted directly from TG 109 principles (II, V, and III) [1].

2. Ethics in the context of the AAPM's Code of Ethics can best be described as [1]:

    A. The local laws governing an individual or group
    B. The principles of conduct guiding an individual or group
    C. A set of beliefs based on majority opinion
    D. A set of beliefs regarding right and wrong

N. Bahar (✉)
Radiation Oncology, St. Peter's Hospital, Albany, NY, USA
e-mail: ninabahar@nyu.edu

M. Heard et al. (eds.), *Absolute Therapeutic Medical Physics Review*,
https://doi.org/10.1007/978-3-031-14671-8_11

Answer: B. The Code of Ethics aims to provide guidelines (as opposed to laws) for ethical conduct and to serve as a resource for the medical physics community, specifically AAPM members [1]. It does not necessarily reflect the majority opinion in AAPM nor is it based on or intended as an individual's moral compass. Ethics can be thought of as principles of conduct for a group, in this case, a professional one; morals tend to come from within and guide an individual's sense of right and wrong; and laws are rules that regulate society, typically in a more absolute sense and under threat of penalty.

3. In TG 109, which of the following is not a principle [1]?

    A. "Members should support the ideals of justice and fairness in the provision of healthcare and allocation of limited healthcare resources."
    B. "Members are professionally responsible and accountable for their practice, attitudes, and actions, including inactions or omissions."
    C. "Members must hold as paramount the best interest of the patient under all circumstances."
    D. "Members must adhere to the legal and regulatory requirements that apply to the practice of their profession."
    E. None of the above.

Answer: A. This response is nearly a direct quote, the exception being the word "should." The AAPM Code of Ethics makes a distinction between "should" and "must" when referring to different guidelines. "Should" applies to scenarios where following a recommendation or guideline is generally advisable with a few possible exceptions. "Must" suggests that following a given recommendation or guideline is necessary to be in line with the Code. Choices B, C, and D are quoted directly from TG 109 principles (X, I, and VIII) [1].

4. The principles in TG 109 are based on values in medical ethics such as [1]:

    A. Beneficence
    B. Amity
    C. Autonomy
    D. Justice
    E. Absolutism

Answer: A, C, and D. These and other key values are given at the beginning of TG 109 after the Table of Contents. While amity (friendship) is nice, it is not among the core values given. Absolutism can refer to a totalitarian style of government or to the notion that standards or values are immutable—neither is applicable in this context [1].

5. In TG 109, which of the following is not a principle [1]?

    A. "Members must support the ideals of justice and fairness in the provision of healthcare and allocation of limited healthcare resources."
    B. "Members must interact in an open, collegial, and respectful manner amongst themselves and in relation to other professionals, including those in training, and safeguard their confidences and privacy."

   C. "Members must strive to continuously maintain and improve their knowledge and skills while encouraging the professional development of their colleagues and of those under their supervision."

   D. "Members must operate within the limits of their knowledge, skills, and available resources in the provision of healthcare. Members must enable practices in which patients are provided the levels of medical physicist expertise and case-specific attention as appropriately supports the modalities of their care."

   E. None of the above.

Answer: E. A–D are quoted directly from TG 109 principles (IX, IV, VI, VII). There are ten principles in total [1].

6. Which of the following is not covered in discussing the environment and ethics of the workplace in TG 109 [1]?

   A. Diversity

   B. Equity

   C. Competence

   D. Conflicts of interest

Answer: B. Diversity is discussed in 3.I.C.a, competence in 3.I.D.h, and conflicts of interest in 3.I.D.j in TG-109 [1].

7. In discussing responsibility to peers and to the profession, TG 109 mentions that members have a responsibility to do everything except [1]:

   A. Build an optimal practice environment

   B. Advance awareness and understanding of what it means to be a medical physicist

   C. Diligently conduct their work

   D. Meet all organizational or institutional demands

Answer: D. TG 109 does not stipulate answer D, whereas A–C are paraphrased from 3.I.A.a. TG 109 does mention that if workplace demands include behavior that is at odds with the principles, then the member should take appropriate action to resolve the conflict [1].

Part II: Clinical Ethics
Unless otherwise specified, select all applicable answers.

1. The department will not provide the resources sufficient to establish a special procedure program that it is seeking to establish. In this type of scenario, TG 109 advises that it is obligatory for the medical physicist to [1]:

   A. Explain the extent of work performed

   B. Resign

   C. Advocate for changes within the department that will afford sufficient resources

   D. Explain the extent of work not able to be performed
   E. Suggest referral of special procedure cases elsewhere

Answer: A, C, D, E. This is discussed in 3.II.C [1].

2. Which types of communication play a role in clinical ethics [1]?

   A. Communication with the public
   B. Communication with other health providers
   C. Communication with caregivers
   D. Communication between managers and subordinates

Answer: A, B, C, and D. A–C are directly addressed in 3.II.A and 3.II.B. D is indirectly addressed in 3.II.C in TG 109 [1].

3. With respect to clinical ethics, members must do the following except [1]:

   A. Engage in continuing education
   B. Regard the patient's best interest with the utmost importance
   C. Respect patient privacy
   D. Regard their employer's interests as paramount

Answer: D. Section 3.II.A of TG 109 addresses continuing education, patient privacy, and holding the interest of the patient as paramount; A–C are paraphrased from this section [1].

Part III: Research Ethics
Unless otherwise specified, select all applicable answers.

1. The use of informed consent puts into practice the guidelines espoused by the:

   A. Belmont Report
   B. Helsinki Declaration
   C. Geneva Convention
   D. Helsinki Accords

Answer: A and B. Both the Belmont Report and the Helsinki Declaration concern human participants in research studies. C and D are notable for other historic reasons but are not relevant in this context [2, 3].

2. Placebo use is considered to be an option if:

   A. There is no proven treatment for a given condition
   B. It is necessary to verify the safety of a given treatment
   C. It is onerous to prove that a new treatment is as effective as an existing one
   D. Patients will not face additional risks or serious harm from a placebo

Answer: A, B, and D. Placebo use is not without its controversy. TG 109 references the Helsinki Declaration, which advises A, B, and D [1, 2].

3. In the context of research with human subjects, beneficence:

   A. Is explicitly identified in the Helsinki Declaration
   B. Is an obligation
   C. Translates to maximizing potential benefits while minimizing potential negative consequences
   D. May require the care provider to balance risks and benefits in considering an intervention

Answer: B, C, and D. TG 109 references the Belmont Report, which advises B, C, and D. D is particularly interesting because, as the Belmont Report observes, in order to avoid or minimize the likelihood of harm, one must know what is harmful. It may be research involving human participants that better informs which interventions can increase the likelihood of harm. This is weighed against the need for intervention and additional measures taken to reduce risk [1, 3].

4. In the context of research ethics with animal participants, the 3 Rs are:

   A. Replacement
   B. Reduction
   C. Refinement
   D. Reparation
   E. Rotation
   F. Repair

Answer: A, B, and C. This is per the Principles of Humane Experimental Technique which TG 109 references. They are viewed as key in improving the welfare of animal research participants. Developed in 1959, the Principles of Humane Experimental Technique is the product of collaboration between animal welfare organizations and the scientific community. Replacement refers to using an insentient participant if one is available in lieu of a sentient participant. Reduction refers to using as few animals as possible in conducting research. Refinement is taking measures to mitigate distress that the animals may experience [1, 4, 5].

5. True or False: Members must declare financial interests when submitting manuscripts or giving presentations even if said interests are tangential to the subject being discussed [1].

Answer: True. TG 109 states this in Sect. 3.III.D.b [1].

6. When submitting work for publication, members must declare which of the following regarding their relationships with business or corporate entities according to TG 109 [1]?

   A. Sponsorship
   B. Travel reimbursement
   C. Bonus incentives
   D. Stock ownership

Answer: A, B, C, and D. TG 109 states this in Sect. 3.III.D.b [1].

Part IV: Education Ethics
Unless otherwise specified, select all applicable answers.

1. In TG 109, educators and trainees both have specific guidelines regarding [1]:

    A. Acknowledgement of work performed by others
    B. Respect
    C. Romantic relations between educators and trainees
    D. Intellectual property

Answer: A, B, and C. Intellectual property is not among the topics covered in guidelines for educators in TG 109. Acknowledging work performed by others occurs in 3.IV.A.b and 3.IV.B.c. Respect comes up in 3.IV.A.b and 3.IV.B.a. Intimate relationships are discussed in 3.IV.A.e and 3.IV.B.d [1].

2. Educators must [1]:

    A. Support trainees in achieving their goals
    B. Promote a safe environment for learning
    C. Share student information with other educators
    D. Make fair evaluations of trainees' efforts

Answer: A, B, and D. There may be situations where it is appropriate or possibly in the student's best interest for educators to share student information; however, the Code of Ethics suggests that confidentiality be maintained appropriately with regard to student information in Sect. 3.IV.A.d. Without further context, it is not clear that sharing information would be appropriate. A, B, and D are addressed in 3.IV.A., 3.IV.A.a, and 3.IV.A.f [1].

Part V: Ethics in Business and Government
Unless otherwise specified, select all applicable answers.

1. In TG 109, "a member who provides a client with domain expertise and advise in exchange for compensation" is a (n) [1]:

    A. Consultant
    B. Contractor
    C. Counsel
    D. Advisor

Answer: A. This is defined in TG 109 Sect. III.V.D [1].

2. In TG 109, "a member who enters into a formal or informal arrangement with a client to provide routine services to the client in exchange for compensation" is a(n) [1]:

    A. Consultant
    B. Contractor

   C. Counsel
   D. Advisor

Answer: B. This is defined in TG 109 Sect. III.V.D [1].

3. As a courtesy of business, vendors may not offer [1]:

   A. Promotional items
   B. Educational items
   C. Modest gifts
   D. Consultation agreements
   E. Grants

Answer: D, E. This is discussed in Sections III.V.B and III.V.C.a in TG 109 [1].

4. True or False: Research grants and sponsorship should be factors in a facility's equipment purchasing decisions [1].

Answer: False. TG 109 elaborates on this in Sections III.V.B.d and III.V.C.c [1].

5. "The practice of contracting for services while simultaneously holding a position as an employee of a different agency or company" is called [1]:

   A. Temping
   B. Locum tenens
   C. Moonlighting
   D. Per diem

Answer: C. This is discussed in Sect. III.V.D.c. of TG 109 [1].

6. True or False: There are no circumstances in which a company may provide a grant to a training institution.

Answer: False. TG 109 recommends reviewing industry codes of ethics, such as ADVAMed, which stipulates that companies may support conference or program attendance costs for trainees in healthcare by providing an educational grant to a training institution provided that the conference or program is third party and the trainees receiving support are chosen by the institution [1, 6].

References

1. Skourou C, Sherouse GW, Bahar N, Bauer LA, Fairobent L, Freedman DJ, Genovese LM, Halvorsen PH, Kirby NA, Mahmood U, Ozturk N, Osterman KS, Serago CF, Svatos MM, Wilson ML. Code of ethics for the American Association of Physicists in Medicine (revised): report of task group 109. Med Phys. 2019;46:e79–93. https://doi.org/10.1002/mp.13351.
2. World medical association. World medical association declaration of Helsinki: ethical principles for medical research involving human subjects. JAMA. 2013;310(20):2191–4. https://

www.wma.net/policies-post/wma-declaration-of-helsinki-ethical-principles-for-medical-research-involving-human-subjects/. Accessed 28 April 2021
3. National Commission for the Protection of Human Subjects of Biomedical and Behavioral Research. The Belmont report: ethical principles and guidelines for the protection of human subjects of research. Office for Human Research Protections; 1979. https://www.hhs.gov/ohrp/regulations-and-policy/belmont-report/read-the-belmont-report/index.html. Accessed 28 April 2021
4. Hubrecht RC, Carter E. The 3Rs and humane experimental technique: implementing change. Animals. 2019;9(10):754. https://doi.org/10.3390/ani9100754.
5. Russell WMS, Burch RL. The principles of humane experimental technique. 1959. https://caat.jhsph.edu/principles/the-principles-of-humane-experimental-technique. Accessed 28 April 2021.
6. Advanced Medical Technology Association. Code of ethics on interactions with health care professionals. 2020. https://www.advamed.org/sites/default/files/resource/advamed-code-of-ethics_2020_july20.pdf. Accessed 28 April 2021.

Chapter 12
Imaging

Vasanthan Sakthivel and Raghavendiran Boopathy

1. Which of the following tube-detector relationships attribute to ring artifacts on the computed tomographic image?

   A. Rotate-translate
   B. Translate-rotate
   C. Rotate-rotate
   D. All of the above

Answer: C
Ring artifact is caused by a miscalibrated or defective detector element, which results in rings centered on the center of rotation. This can often be fixed by recalibrating the detector. A scanner with solid-state detectors, where all the detectors are separate entities, is in principle more susceptible to ring artifacts [1].

2. How will you define "pitch" for a helical CT?

   A. Table movement in 360°/beam width
   B. Projections at neighboring scan-axis positions
   C. Reconstructed slice thickness
   D. A plane through the body perpendicular to the scan-axis

V. Sakthivel (✉)
Department of Radiation Oncology, Cancer Specialists of North Florida,
Jacksonville, FL, USA
e-mail: Vasanthan.sakthivel@csnf.us

R. Boopathy
Department of Radiation Oncology, OU Stephenson Cancer Center, University of Oklahoma
Health Science Center, Oklahoma City, OK, USA
e-mail: raghavendiran-boopathy@ouhsc.edu

© The Author(s), under exclusive license to Springer Nature Switzerland AG 2022   103
M. Heard et al. (eds.), *Absolute Therapeutic Medical Physics Review*,
https://doi.org/10.1007/978-3-031-14671-8_12

Answer: A

Pitch (P) is a term used in helical CT. It has two terminologies depending on the scanner type (single or multislice). Choice of pitch affects both image quality and patient dose.

Single-slice CT:

The term detector pitch is used and is defined as the table distance traveled in one 360° gantry rotation divided by beam collimation.

Multislice CT:

Beam pitch is defined as the table distance traveled in one 360° gantry rotation divided by the total thickness of all simultaneously acquired slices [2].

3. If only the patient size changes, provided that we keep all other scan parameters the same, what happens to the radiation dose delivered?

   A. Increasing patient size causes decreased dose.
   B. Increasing patient size causes an increased dose.
   C. No relation between patient size and dose.
   D. Increasing patient size does not change the dose.

Answer: B

CTDI (measured in mGy) is a standardized measure of the radiation dose output of a CT scanner, which allows the user to compare the radiation output of different CT scanners.

CTDI100 (mGy) is a linear measure of dose distribution over a pencil ionization chamber and hence does not take into consideration the topographical variation of a human body and is therefore not in clinical use.

CTDIw (mGy) is closer to the human dose profile as compared with the CTDI100:

2/3 CTDI100 (periphery) + 1/3 CTDI100 (center)

CTDIvol (mGy) is obtained by dividing CTDIw by pitch factor.

Another commonly used index is the dose-length product (DLP), which factors in the length of the scan to show the overall dose output.

DLP (mGy*cm) = CTDIvol × scan length.

CTDI is a measure of patient dose. The concerning X-ray tube's output must be an increased actual dose to any given patient, which is directly dependent on the size and shape of the patient. To achieve similar image quality, the scanner output (CTDIvol) should be increased by about a factor of two as patient size changes from a typical adult abdomen (lateral dimension, 35–40 cm) to an obese adult abdomen (lateral width, 45–50 cm) [3, 4].

4. What does the beam hardening in X-ray imaging means:

   A. Decrease in a photon's energy as it is scattered by a dense material
   B. Increased average X-ray energy as a beam passes through a dense material
   C. Decreased average X-ray energy as a beam passes through a dense material
   D. Increase in a photon's energy as it is scattered by a dense material

Answer: B

Beam hardening is the occurrence that happens when an X-ray beam composed of polychromatic energies passes through an object, resulting in selective attenuation of lower energy photons. The effect is similar to a high-pass filter, in that only higher energy photons are left to contribute to the beam and thus the mean energy of the beam is increased [5].

5. While performing an ultrasound scan, the technician notices that the echoes at a depth of 5–7 cm appear relatively weak. What parameter needs tweaking to increase the brightness of the image?

   A. Frequency
   B. Focusing
   C. TGC
   D. Beam intensity

Answer: C

The time-gain compensation (TGC) is the control that can be used to selectively amplify the echoes from any specific depth. This adjustment is accomplished by TGC controls that permit the user to selectively amplify the signals from posterior structures or to suppress the signals from anterior tissues, thereby compensating for tissue attenuation [6].

6. Which of the following increases the signal-to-noise ratio of a CT image?

   A. Decreased aperture size
   B. Decreased milliampere seconds (mAs)
   C. Increased aperture size
   D. Increased filtration

Answer: C

SNR is a generic term that, in radiology, is a measure of true signal to noise.

In CT, the signal-to-noise ratio is determined by:

mAs: greater mAs increase SNR.

The mAs or the dose of a CT scan has a direct relationship with the number of photons utilized in the examination.

$2\times$ mAs = 40% increased SNR.

Slice thickness: thicker slices increase SNR.

Scan slice thickness is often very thin at 1 mm. However, viewing is thicker, at 4 mm, averaging 4.

1 mm Slices into one thick 4 mm will increase the signal by $4\times$, but the noise will increase $\sqrt{4} = 2\times$.

Patient size: larger patients reduce SNR [7].

7. What factor does the subject contrast for a CT image depends on?

    A. kV
    B. Focal spot
    C. mA
    D. Collimation

Answer: A

Subject contrast is the ratio of transmitted radiation intensities and is, thus, dependent on the absorption differences in the subject, photon energy, and scattered radiation. Low-kilovoltage photons will generally result in a radiograph with high contrast. mA will impact the signal-to-noise ratio. The focal spot will impact image sharpness with a larger penumbra associated with a larger focal spot [3].

8. What image quality parameter is being evaluated in this image?

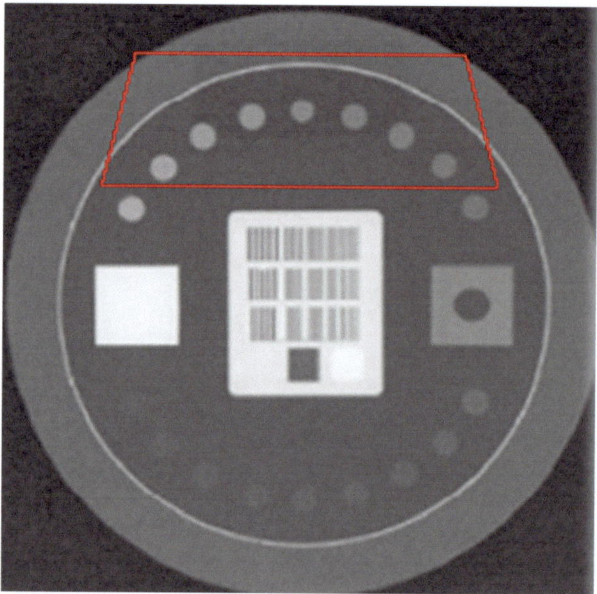

    A. Spatial resolution
    B. Noise
    C. Contrast
    D. Uniformity

Answer: C

The test object shown is composed of discs of varying densities and is used to evaluate low-contrast detectability. Contrast refers to the ability to distinguish subtle differences in material composition [8].

9. The shown radiographic image's quality is affected by:

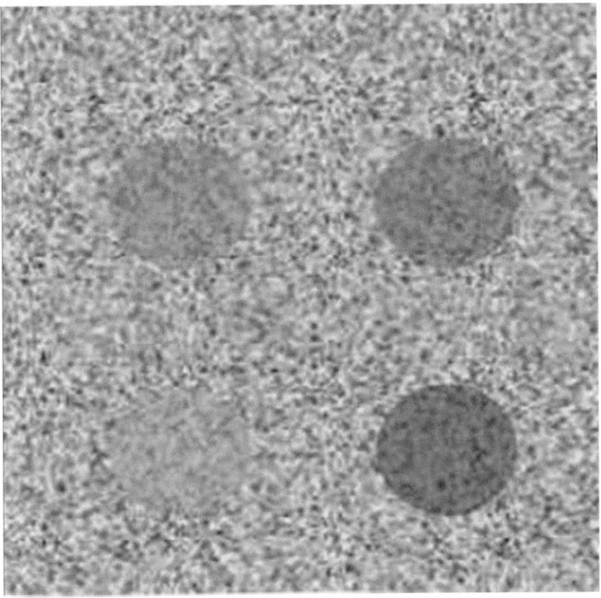

   (i) Quantum mottle
  (ii) Unsharpness
 (iii) Resolution
 (iv) Contrast

    A. All the above
    B. None of the above
    C. i, ii, and iii
    D. i, ii, and iv

Answer: C
Quantum mottle is the statistical fluctuation of the number of photons absorbed by the intensifying screens to form the light image on the film. Quantum mottle artifact is when the X-rays are produced but are not produced uniformly. Quantum mottle noise is a result of an inefficient number of photons reaching the imaging plate due to an error in the preset exposure factors (mAs and kVp). Quantum mottle will cause fluctuation in image densities, resulting in images with a grainy appearance. This in turn creates unsharpness and poor resolution in the obtained image.

10. What is the CT Hounsfield unit of water?

    A. 500
    B. −1000
    C. 0
    D. −700

Answer: C
The tolerance for the CT Hounsfield unit of water is 0 ± 5 HU. A Hounsfield unit value of -1000 is typically representative of air, -700 is typically representative of the lung, and 500 is typically representative of bone [9].

11. A modulation transfer function equal to 1.0 at a particular spatial frequency (f) means:

    A. An electronic circuit utilized with automatic brightness control is set to perfect linearity.
    B. Energy transferred to tissue is equal to that of the incident radiation.
    C. The diagnostic image perfectly reproduces contrast variations in the object radiographed at that spatial frequency.
    D. High-voltage ripple transmitted from mains fluctuations is 1%.

Answer: C
The modulation transfer function (MTF) determines how much contrast in the original object is maintained by the detector. In other words, it characterizes how faithfully the spatial frequency content of the object gets transferred to the image. The modulation transfer function of an image intensifier is a measure of the output amplitude of dark and light lines on the display for a given level of input from lines presented to the photocathode at different resolutions. It is usually given as a percentage at a given frequency (spacing) of light and dark lines [10].

12. Quality of digitally reconstructed radiographs depends upon:

    (i) Pitch
    (ii) Slice thickness
    (iii) FOV
    (iv) Mode

        A. All the above
        B. None of the above
        C. i, ii, and iii
        D. i, ii, and iv

Answer: C
Digitally reconstructed radiographs (DRRs) are computed transmission images of the patient's anatomy in the beam's-eye view calculated from computed tomography (CT).

Dataset onto a planar view: They are generated from the same CT datasets used for treatment planning and provide a reference image for the verification of patient position during radiotherapy treatment. DRRs are generated from CT images; the image quality may be affected by the CT scanning mode (axial or helical), parameters such as scanning pitch and reconstruction pitch, as well as the number of virtual X-ray lines.

Projected from the virtual source in the calculation of the DRR: Helical scanning has several advantages over axial scanning such as fast volumetric data acquisition lower dose to the patient and reduced beam-hardening streak artifacts. However, unlike axial scanning, longitudinal slice positions in helical scanning are reconstructed using a variety of interpolation schemes, which can lead to loss of spatial resolution in the longitudinal direction. Aliasing in helical scanning at pitches greater than 1 also influences the spatial resolution of multiplanar images. Therefore, the quality of DRRs generated from helical scanning data can potentially be influenced by the helical pitch, slice thickness, and scanned field of view (FOV) [11].

13. What are the disadvantages of double-exposure port film?

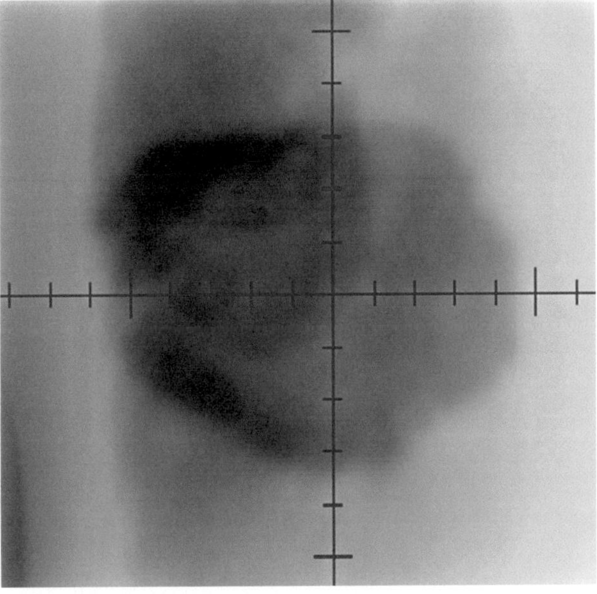

  (i)   Additional anatomical landmarks and details
 (ii)   Probability of stochastic effect
(iii)   Imaging dose cannot be accounted with treatment dose
 (iv)   Image blurring

    A. i and ii
    B. ii and iii
    C. All of the above
    D. None of the above

Answer: B

Localization radiograph is produced by a sequence of two exposures, first to a shaped treatment field and then to a larger rectangular field. The resulting image serves to locate the treatment field borders with respect to the patient's anatomy.

Portal images are acquired either by single-exposure technique or by double-exposure technique. In a single-exposure technique, the treatment beam is used to image the region to be treated when adequate landmarks are available for verification within this region. The dose delivered during this single-exposure portal imaging is usually adjusted from the treatment dose, and this does not deliver the dose to normal tissues. When adequate landmarks are not available within the treatment region, a double-exposure technique is used. In this double-exposure technique, a field at least 5 cm larger than the area to be treated is also imaged in addition to the region to be treated. This results in delivering dose outside the tumor volume and thus increases the probability of stochastic effect. The risk of cancer induction is additive, and the concomitant dose from the double-exposure portal images adds to the dose received by the patient due to leakage and scatter radiation [12].

14. Identify and order differently weighted MRI images:

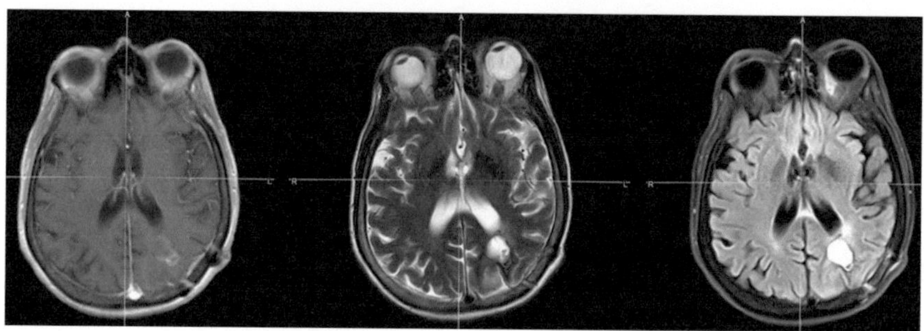

    A. T1, T2, FLAIR
    B. T2, T1, FLAIR
    C. FLAIR, T2, T1
    D. T1, FLAIR, T2

Answer: A

CSF appears dark in a T1-weighted and FLAIR image and bright in a T2-weighted image. White matter appears light in T1 and dark gray in T2 and FLAIR. Fat appears bright in T1 and light in T2 and FLAIR images [13].

15. Which of the following is used for attenuation correction on PET/CT scanners?

    A. PET emission scan only
    B. The CT portion of the exam and associated attenuation values
    C. K-space values
    D. A transmission scan is performed after the CT

Answer: B

PET image data scanned with PET/CT scanners is attenuation corrected using data from computed tomography (CT). Although CT-based attenuation correction is generally reliable, patient movement and remains of CT contrast media may cause quantitative errors and image artifacts. Respiratory movement causes differences in the CT-based attenuation map and the actual attenuation during PET scan, which may cause bias.

It is important to note that attenuation-corrected dataset can introduce several artifacts like contrast-based artifacts, implanted device attenuation artifacts, and diaphragmatic respiratory artifacts. Hence, it is imperative to become familiar with the common and atypical appearances of the most common artifacts and be diligent about inspecting the non-AC PET images when an AC artifact is suspected [14, 15].

16. Compared with a 1.5-T MRI scanner, what is the primary advantage of a 3-T MRI scanner?

    A. Lower price
    B. Higher signal-to-noise ratio (SNR)
    C. Lower energy deposition in tissues
    D. Increased safety for patients with medical devices or implants

Answer: B

The amount of energy deposited in tissues by radiofrequency pulses (specific absorption rate [SAR]) is proportional to the square of magnetic field strength and, thus, higher at high B fields [16].

17. In stereotactic treatment planning, the MRI data are typically fused with CT images. What is an advantage of using MRI compared with CT?

    A. Short acquisition time
    B. Higher spatial resolution
    C. Superior soft-tissue contrast
    D. Accurate electron density

Answer: C

CT is commonly used in radiation therapy planning because it provides superior spatial accuracy and electron density information necessary for heterogeneity corrections. MRI provides superior soft-tissue contrast and visualization of tumor invasion of surrounding soft tissues. A disadvantage of MRI is its inferior spatial accuracy and lack of electron density information. Another MRI disadvantage is the inability to differentiate between bone and air. Also, long scan times and absence of immobilization make MRI susceptible to patient motion artifacts [17].

18. What CT image set is best suited for treatment planning dose calculations for a lung lesion to be treated with radiotherapy?

    A. Minimum intensity projection CT
    B. Maximum intensity projection CT
    C. Free-breathing CT
    D. Average intensity projection CT

Answer: D

An average intensity projection (AIP) dataset derived from a 4D CT scan has been shown to be most useful for both treatment planning dose calculations and verification of tumor location via image-guided radiation therapy. Free-breathing datasets are prone to significant image artifacts, and MIP and MinIP datasets may not accurately represent the target volume, especially when the target is close to more dense tissues [18].

19. What registration technique involves a transformation to preserve the distance between all points in the image and can include translation and rotations in all directions? Registration QA:

    A. Affine
    B. Rigid
    C. Deformable
    D. Fusion

Answer: B

Rigid registration involves transformation to preserve the distance between all points in the image and can include translation and rotations in all directions. Image fusion is a technique where the mapped data from the moving dataset are combined with the stationary dataset. Deformable registration transformation can also be spatially variant where the number of degrees of freedom can be as large as three times the number of voxels in the source dataset. It warps a moving image via a deformable field to align with a target image and defines the motion of each voxel from the moving image to the target image. Affine registration includes transformation from rigid registration and adds additional transformation of scaling, shearing, and plane reflection. The distance between all points is not maintained as in rigid registration, but the parallel lines remain parallel after transformation [19].

20. What is the main advantage of MV-CBCT over KV-CBCT?

    A. Low imaging dose for the patient
    B. Fewer artifacts due to metal implants
    C. Better visualization of soft tissue
    D. Improved image contrast

Answer: B

Due to the dominant interaction of the Compton effect, MV cone-beam CT is less susceptible to metallic artifacts and produces images with less contrast because it depends more on electron density than atomic number. The decrease, in contrast, impacts the ability to visualize soft tissue. CT numbers created with MV cone-beam CT also have a direct correlation with electron density with the therapeutic beam, and no extrapolation is needed to calculate the imaging dose [20].

21. What is the most appropriate imaging modality to guide the placement of low-dose-rate brachytherapy seeds in the prostate?

    A. MRI
    B. CT

C. Ultrasound
D. Fluoroscopy

Answer: C

Transrectal ultrasound is the preferred imaging modality when performing permanent prostate brachytherapy. Fluoroscopy is oftentimes used to determine needle and/or distal seed extent relative to the bladder, which is usually filled with contrast but is not used for the 3D positioning of seeds [21].

22. What is an advantage of filtered back projection compared with iterative reconstruction methods?

    A. Filtered back projection is less sensitive to missing information.
    B. Filtered back projection is faster.
    C. Filtered back projection results in a lower dose to the patient for the same signal-to-noise ratio.
    D. Filtered back projection can be used with all 3D modalities.

Answer: B

Filtered back projection was the primary reconstruction method for CT imaging, including that used by CT simulators for therapy planning. Iterative methods are more complicated and correspondingly more computationally intensive; however, as methods improved and computers became faster, they became practical for the clinic. For the same patient dose, iterative methods offer better signal-to-noise ratios than filtered back projection. Additionally, iterative methods are less sensitive to missing information and, therefore, less prone to metal artifacts [22].

23. Which image registration method tries to align voxels whose values have common probabilities of being present in their respective image sets?

    A. Mutual information
    B. Sum of squared differences
    C. Correlation coefficient
    D. Mean squared difference

Answer: A

Voxel-based registration methods include the sum of the squared difference (SOSD), mean squared difference (MSD), correlation coefficient (CC), and mutual information. SOSD and MSD assume an equivalent relationship between voxel intensities in the two images and only really work when both image sets are in the same modality. The normalized CC metric measures the similarity in the image signal (so should be used with images that have similar voxel intensities for similar organs) and maximizes the intensity product with the assumption that there is a linear relationship between the intensity values in each image; it can handle differences in image contrast better than the first two metrics, but requires the voxel intensities to be correlated. Only mutual information, which assumes a statistical relationship between the voxel intensities with no dependence on absolute intensity values, is appropriate with two very different modalities like CT and MRI [19].

24. Resolution of the imaging system is a characteristic that is directly related to:

  (i) Image unsharpness
  (ii) Visibility of anatomical detail
  (iii) Image noise
  (iv) Image blurring

    A. i
    B. i, ii, and iv
    C. iii and iv
    D. None of the above

Answer: B

Image resolution is the detail an image holds. Higher resolution means more image detail. Resolution is the ability of an imaging system to reproduce a sharp edge that is present in the object. The key factors that influence the sharpness of an image relate to the size of the source of X-rays, the physical characteristics of the X-ray detector system, and the presence of any motion blur because of the finite duration of all radiographic exposures. Blurring refers to the distortion of the definition of objects in an image, resulting in poor spatial resolution. Blurring produces image unsharpness, decreases the visibility of anatomical detail, and reduces resolution if the resolution is being measured [6].

References

1. Martz HE, Logan CM, Schneberk DJ, Shull PJ. X-ray imaging fundamentals, industrial techniques and applications 2017.
2. Allisy-Roberts PJ, Williams J. Farr's physics for medical imaging. 2nd ed; 2008.
3. Bushberg JT, Seibert JA, Leidholdt EM Jr, Boone JM. The essential physics of medical imaging. Lippincott Williams & Wilkins; 2011.
4. Wilting JE, et al. A rational approach to dose reduction in CT: individualized scan protocols. Eur Radiol. 2001;11(12):2627–32.
5. Andreas Maier, Stefan Steidl, Vincent Christlein, Joachim Hornegger. Medical imaging systems an introductory guide 2018.
6. Sprawls P. The physical principles of medical imaging. 2nd ed; 1995.
7. Seeram E. Computed tomography: physical principles, clinical applications, and quality control. 4th ed; 2015.
8. Fontenot JD, Hassaan A, et al. AAPM medical physics practice guideline 2.A: commissioning and quality assurance of X-ray-based image-guided radiotherapy systems. J Appl Clin Med Phys. 2014;15(1):4528.
9. Report No. 083–Quality assurance for computed-tomography simulators and the computed-tomography-simulation process: report of the AAPM radiation therapy committee task group no. 66. Med Phys. 2003;30:2762–92.
10. Samei E, et al. Comparison of edge analysis techniques for the determination of the MTF of digital radiographic systems. Phys Med Biol. 2005;50(15):3613–25.
11. Michael Goitein. Radiation oncology: a physicist's—eye view. 2008.

12. Ravindran P. Dose optimisation during imaging in radiotherapy. Biomed Imaging Interv J. 2007;3(2):e23.
13. https://case.edu/med/neurology/NR/MRI%20Basics.htm.
14. Carrio I, Ros P, editors. PET/MRI–methodology and clinical applications. Springer; 2014.
15. Blodgett TM, et al. PET/CT artifacts. Clin Imaging. 2011;35(1):49–63.
16. Zhou J, et al. Imaging techniques in stereotactic radiosurgery. In: Chin L, Regine W, editors. Principles and practice of stereotactic radiosurgery. New York: Springer; 2015. https://doi.org/10.1007/978-1-4614-8363-2_2.
17. Katsuragawa, S., Doi, K. Hendee's physics of medical imaging, fifth edition, by Ehsan Samei and Donald JP. Radiol Phys Technol. 2019;12: 438–439.
18. Tian Y, Wang Z, Ge H, Zhang T, Cai J, Kelsey C, Yoo D, Yin FF. Dosimetric comparison of treatment plans based on free breathing, maximum, and average intensity projection CTs for lung cancer SBRT. Med Phys. 2012;39(5):2754–60. https://doi.org/10.1118/1.4705353.
19. Report No. 132—Use of image registration and fusion algorithms and techniques in radiotherapy: report of the AAPM radiation therapy committee task group no. 132. 2017.
20. Khan FM. Khan's the physics of radiation therapy. Fifth ed. Lippincott Williams & Wilkins; 2014.
21. Davis BJ, et al. American Brachytherapy Society. American Brachytherapy Society consensus guidelines for transrectal ultrasound-guided permanent prostate brachytherapy. Brachytherapy. 2012;11(1):6–19.
22. Klink T, Obmann V, Heverhagen J, Stork A, Adam G, Begemann P. Reducing CT radiation dose with iterative reconstruction algorithms: the influence of scan and reconstruction parameters on image quality and CTDIvol. Eur J Radiol. 2014;83(9):1645–54. https://doi.org/10.1016/j.ejrad.2014.05.033.

Chapter 13
Treatment Planning Systems

Brandon Merz

1. A structure with volume 99 cc is defined in TPS A. The structures are exported from TPS A to TPS B. Once imported into TPS B, the volume appears as 100 cc. What would be a likely cause of this discrepancy?

   A. Structure geometric representation in database
   B. TPS B has a bug in the method which reports structure volume
   C. Storing structure data in DICOM format for exchange
   D. Insufficient memory on workstations

Answer: The answers are A and C. The structure will have an internal geometric representation in TPS A, which will be stored in a database/memory. In order to transport this to another program, it will need to essentially be binned into a file format for transfer. The DICOM standard defines the format for transfer [1]. TPS B will then read this binned information at import and use it to create an internal representation. The format for structure representation in TPS B is most likely defined differently from TPS A and is also being constructed from an intermediate exchange, greatly increasing the chances for minor discrepancies in reported structure volume [2].

2. Which of the following resolution considerations play the most significant role with respect to target margins?

   A. Voxel size
   B. Dose grid resolution
   C. Slice thickness

B. Merz (✉)
Radiation Medicine, Oregon Health and Science University, Portland, OR, USA
e-mail: merzbr@ohsu.edu

M. Heard et al. (eds.), *Absolute Therapeutic Medical Physics Review*,
https://doi.org/10.1007/978-3-031-14671-8_13

Answer: The answer is C. Voxel sizes are generally well below 1 mm. Dose grid can be selected to very high resolution and rarely influences the decision for structure margin. The slice thickness in some circumstances (volume averaging for example) can motivate discussion and selection of target margins [3].

3. You experience an error upon trying to import a structure to your TPS. Where would be an appropriate first place to check for answers?

    A. NEMA standard definitions
    B. Vendor's DICOM conformance statement
    C. Vendor support desk
    D. DICOM image/file-viewing program

Answer: The answer is B. Physicist should have knowledge of the capabilities and limitations of a TPS. This includes monitoring vendor documentation and customer technical bulletins and distributing relevant information to the department. A reasonable first step would be to check if the file format is supported by the TPS [4]. From there, the file could be briefly inspected with a DICOM file-viewing utility. If still unresolved, vendor support can be utilized after gathering information to accurately describe the issue.

4. Which of the following could degrade modulated plan quality by needlessly consuming scarce optimization resources?

    A. Automated planning algorithms
    B. Choice of structure priorities
    C. Regions with steep dose gradients
    D. Normal and target structure overlap in patient model

Answer: The answers are B and D. If a lower importance structure is given a higher priority weight relative to a higher importance structure, the penalty score during optimization will result in it getting more of the optimization resources.

Overlap in structures with competing objectives is also a common issue, which will consume optimization resources unnecessarily and thus lower the chances of finding an optimal plan. Longer than usual optimization times can be a symptom of this problem [5].

5. What is the result of increasing the priority weighting for a single objective function in an optimization problem?

    A. The optimization problem will require more time to converge on a satisfactory result.
    B. Improved TCP/NTCP.
    C. The significance of the penalty score will increase for that structure relative to others.
    D. Improved ability to find optimal plan quality with respect to all objective functions.

Answer: The answer is C. Each iteration of optimization will sum the penalties for the structures involved in the optimization problem [6]. The priority weighting can cause a saturation of focus on certain objectives if not used judiciously.

6. What is the clinical risk of updating the Windows version to a non-FDA-approved combination of OS and TPS?

   A. GPU driver will no longer be supported in new Windows version, forcing the use of CPU for calculations.
   B. Changes to system methods used in software via .dll files could result in undetected changes in TPS behavior.
   C. Regulatory action against the center for using a non-approved medical device on patients.
   D. Invalidation of TPS contract with vendor.

Answer: The answer is B. It is important to distinguish between risk and inconvenience. If the underlying system dependencies change, then the TPS could perform in ways that have not been tested or characterized. A search engine can be used to find useful resources on medical device classification as well as what is happening during a Windows upgrade [7]. Coordination with IT departments is often necessary to keep them from updating the operating system to non-approved versions. Further, even the installation of third-party software on planning computers could compromise resources used by the TPS. One should always consult with the vendor and review documentation prior to upgrading.

7. Incorporating the concept of robustness into optimization and evaluation could reduce the need for which of the following?

   A. ITV
   B. PTV
   C. CTV
   D. GTV

Answer: The answer is B. The internal target volume (ITV), gross tumor volume (GTV), and clinical target volume (CTV) are anatomical volumes that encompass the tumor and subclinical disease [8]. The planning target volume (PTV) is a geometric concept to account for uncertainty and ensure that adequate dose is delivered to the other volumes.

   Robust planning and optimization attempt to model target coverage response to motion [9]. The concept is used in proton treatment planning to account for setup and range uncertainty.

8. A vendor specifies that data below 2 cm × 2 cm will not be used in the fit of a beam model. It would be correct to conclude that this TPS is not suitable for IMRT/VMAT.

   A. True
   B. False

Answer: The answer is B. The dose calculation algorithm will still look up output factor information for areas smaller than 2×2 cm^2 regardless of the data the model fit uses during the beam commissioning process. It is advisable to ensure that the data is consistent with published values and perform an independent end-to-end test such as those provided by IROC-Houston [10].

9. What would be an advantage of highly accurate intermediate dose calculation algorithm?

    A. Faster optimization times
    B. Better agreement between optimal fluence and deliverable plan
    C. Improved optimization search
    D. More accurate DVH statistics

Answer: The answers are A and B. The intermediate dose calculation runs a simplified dose calculation, which helps to point the subsequent optimizations in the right direction by applying MLC and jaw characteristics to the optimal fluence [11]. Without an intermediate dose calculation, it would be possible for the isodose lines to reflect excellent, but ultimately undeliverable, target coverage. Furthermore, utilization of the intermediate dose calculation algorithm has demonstrated improvements in planning efficiency by reducing the number of iterations needed to meet the optimization objections and optimization time.

10. A newly released dose calculation algorithm delivers 100% Γ (0.5 mm, 0.5%) over the entire model space (30 cm $\times$ 30 cm $\times$ 30 cm), for all machines, even with highly heterogeneous interfaces. What might be a reason this algorithm would not be immediately adopted?

    A. Insufficient literature review.
    B. The isodose lines look different from the previous TPS.
    C. Insufficient number of physics FTEs.
    D. An end-to-end test has not been performed using the new model.

Answer: The answers are B and D. While answer B is a bit comical, there is the possibility that dosimetrists and physicians will need to adapt and be convinced that what they are seeing is accurate. IROC-Houston has demonstrated discrepancies in TPS calculations, and measurements are not uncommon in clinical practice [12]. An independent end-to-end phantom irradiation is one way to prevent such errors.

11. Which structures should be incorporated into the treatment plan dose calculation?

    A. Bolus
    B. Support structure
    C. Couch
    D. Body

Answer: The answers are A, B, C, and D. Inaccuracies in the contouring of these structures or excluding them in the dose calculation can lead to inaccuracies in the final dose calculation [13, 14]. Consult treatment planning system documentation to verify the dose calculation domain.

12. Which of the following should be performed as part of a TPS QA program?

    A. Monitor vendor literature and technical safety notices for known issues and utilize file integrity checksums.
    B. Daily end-to-end testing of image transfer between CT, TPS, and machine imaging software.
    C. Ensure that second calc check software is independent of primary TPS model.
    D. On a routine basis, representative plans for each configured photon beam should be recalculated.

Answer: The answers are A and D. Monitoring vendor literature and communicating relevant information to members of the radiation oncology team are integral parts of a QA program [15]. According to MPPG 5a [16], physicist should recalculate representative plans. These recalculations should agree with the reference dose calculation to within 1%/1 mm. Answer "B" would be time prohibitive unless the task could be automated.

13. How will increasing the dose grid from a setting of 0.1 mm to 0.2 mm affect the value of a max dose point?

    A. Increase
    B. Decrease
    C. No change is likely

Answer: The answer is B. The maximum dose voxel will now be grouped with more voxels which, by definition, were previously not the maximum dose voxel. Averaging the max voxel with any other voxels will result in a lowering of the max dose point. This can become especially important in plans with steep dose gradients, like stereotactic treatment plans [17, 18].

14. A CT scanner at a satellite was used to simulate a patient for a palliative treatment; however, the CT calibration curve of the scanner at the primary facility was assigned in the TPS system for the patient plan. What is the effect of using a different calibration curve for a photon treatment plan and for an electron treatment plan?

    A. Dose stays the same; D80 is unchanged.
    B. Dose stays the same; D80 is shifted.
    C. Dose errors are a function of tissue thickness and beam energy; D80 is shifted.
    D. Dose errors are only a function of beam energy; D80 is shifted.

Answer: The answer is C. Kilby et al. [19] demonstrated that errors in electron density can cause dose errors >2%, as a function of tissue thickness and photon energy. In addition, errors in density will cause a shift in the dose of 2 mm for an electron plan. Therefore, it is important to perform routine quality assurance of the CT scanner and select the correct CT calibration curve in the TPS.

References

1. Law MY, Liu B. Informatics in radiology: DICOM-RT and its utilization in radiation therapy. Radiographics. 2009;29(3):655–67. https://doi.org/10.1148/rg.293075172.
2. Kim JI, Han JH, Choi CH, An HJ, Wu HG, Park JM. Discrepancies in dose-volume histograms generated from different treatment planning systems. J. Radiat. Prot. Res. 2018;43(2):59–65. https://doi.org/10.14407/jrpr.2018.43.2.59.
3. Srivastava SP, Cheng CW, Das IJ. The effect of slice thickness on target and organs at risk volumes, dosimetric coverage and radiobiological impact in IMRT planning. Clin Transl Oncol. 2016;18(5):469–79. https://doi.org/10.1007/s12094-015-1390-z.
4. Starkschall G, Siochi RAC, editors. Informatics in radiation oncology. 1st ed. CRC Press; 2013. https://doi.org/10.1201/b15508.
5. Wu Q, Mohan R. Algorithms and functionality of an intensity modulated radiotherapy optimization system. Med Phys. 2000;27:701–11.
6. Wilkens J, Alaly J, Zakarian K, Thorstad W, Deasy J. IMRT treatment planning based on prioritizing prescription goals. Phys Med Biol. 2007;52:1675–92. https://doi.org/10.1088/0031-9155/52/6/009.
7. https://www.fda.gov/medical-devices/software-medical-device-samd/what-are-examples-software-medical-device
8. Landberg T, Chavaudra J, Dobbs J, Gerard JP, Hanks G, Horiot JC, Johansson KA, Möller T, Purdy J, Suntharalingam N, Svensson H. Report 62. J ICRU. 1999;32(1) https://doi.org/10.1093/jicru/os32.1.Report62.
9. Chang Y, Xiao F, Quan H, et al. Evaluation of OAR dose sparing and plan robustness of beam-specific PTV in lung cancer IMRT treatment. Radiat Oncol. 2020;15:241. https://doi.org/10.1186/s13014-020-01686-1.
10. Lechner W, Wesolowska P, Azangwe G, Arib M, Alves VGL, Suming L, Ekendahl D, Bulski W, Samper JLA, Vinatha SP, Siri S, Tomsej M, Tenhunen M, Povall J, Kry SF, Followill DS, Thwaites DI, Georg D, Izewska J. A multinational audit of small field output factors calculated by treatment planning systems used in radiotherapy. Phys Imaging Radiat Oncol. 2018;5:58–63. https://doi.org/10.1016/j.phro.2018.02.005.
11. Li Y, Rodrigues A, Li T, Yuan L, Yin F-F, Wu Q. Impact of dose calculation accuracy during optimization on lung IMRT plan quality. J Appl Clin Med Phys. 2015;16:5137. https://doi.org/10.1120/jacmp.v16i1.5137.
12. Kerns JR, Stingo F, Followill DS, Howell RM, Melancon A, Kry SF. Treatment planning system calculation errors are present in most imaging and radiation oncology Core-Houston phantom failures. Int J Radiat Oncol Biol Phys. 2017;98(5):1197–203. https://doi.org/10.1016/j.ijrobp.2017.03.049.
13. Olch AJ, Gerig L, Li H, Mihaylov I, Morgan A. Dosimetric effects caused by couch tops and immobilization devices: report of AAPM task group 176. Med Phys. 2014;41:061501. https://doi.org/10.1118/1.4876299.

14. Wang L, Cmelak AJ, Ding GX. A simple technique to improve calculated skin dose accuracy in a commercial treatment planning system. J Appl Clin Med Phys. 2018;19:191–7. https://doi.org/10.1002/acm2.12275.
15. Fraass B, Doppke K, Hunt M, et al. American Association of Physicists in Medicine radiation therapy committee task group 53: quality assurance for clinical radiotherapy treatment planning. Med Phys. 1998;25(10):1773–829.
16. Smilowitz JB, Das IJ, Feygelman V, Fraass BA, Kry SF, Marshall IR, Mihailidis DN, Ouhib Z, Ritter T, Snyder MG, Fairobent L. AAPM medical physics practice guideline 5.A.: commissioning and QA of treatment planning dose calculations—megavoltage photon and electron beams. J Appl Clin Med Phys. 2015;16:14–34. https://doi.org/10.1120/jacmp.v16i5.5768.
17. Chung H, Jin H, Palta J, Suh T-S, Kim S. Dose variations with varying calculation grid size in head and neck IMRT. Phys Med Biol. 2006;51:4841–56. https://doi.org/10.1088/0031-9155/51/19/008.
18. Rosewall T, Kong V, Heaton R, Currie G, Milosevic M, Wheat J. The effect of dose grid resolution on dose volume histograms for slender organs at risk during pelvic intensity-modulated radiotherapy. J Med Imaging Radiat Sci. 2014;45(3):204–9.
19. Kilby W, Sage J, Rabett V. Tolerance levels for quality assurance of electron density values generated from CT in radiotherapy treatment planning. Phys Med Biol. 2002;47(9):1485–92. https://doi.org/10.1088/0031-9155/47/9/304.

Index